Godfrey C. Onwubolu and Donald Davendra (Eds.)

Differential Evolution: A Handbook for Global Permutation-Based
Combinatorial Optimization

Studies in Computational Intelligence, Volume 175

Editor-in-Chief
Prof. Janusz Kacprzyk
Systems Research Institute
Polish Academy of Sciences
ul. Newelska 6
01-447 Warsaw
Poland
E-mail: kacprzyk@ibspan.waw.pl

Further volumes of this series can be found on our homepage: springer.com

Vol. 153. Carlos Cotta and Jano van Hemert (Eds.)
Recent Advances in Evolutionary Computation for Combinatorial Optimization, 2008
ISBN 978-3-540-70806-3

Vol. 154. Oscar Castillo, Patricia Melin, Janusz Kacprzyk and Witold Pedrycz (Eds.)
Soft Computing for Hybrid Intelligent Systems, 2008
ISBN 978-3-540-70811-7

Vol. 155. Hamid R. Tizhoosh and M. Ventresca (Eds.)
Oppositional Concepts in Computational Intelligence, 2008
ISBN 978-3-540-70826-1

Vol. 156. Dawn E. Holmes and Lakhmi C. Jain (Eds.)
Innovations in Bayesian Networks, 2008
ISBN 978-3-540-85065-6

Vol. 157. Ying-ping Chen and Meng-Hiot Lim (Eds.)
Linkage in Evolutionary Computation, 2008
ISBN 978-3-540-85067-0

Vol. 158. Marina Gavrilova (Ed.)
Generalized Voronoi Diagram: A Geometry-Based Approach to Computational Intelligence, 2009
ISBN 978-3-540-85125-7

Vol. 159. Dimitri Plemenos and Georgios Miaoulis (Eds.)
Artificial Intelligence Techniques for Computer Graphics, 2009
ISBN 978-3-540-85127-1

Vol. 160. P. Rajasekaran and Vasantha Kalyani David
Pattern Recognition using Neural and Functional Networks, 2009
ISBN 978-3-540-85129-5

Vol. 161. Francisco Baptista Pereira and Jorge Tavares (Eds.)
Bio-inspired Algorithms for the Vehicle Routing Problem, 2009
ISBN 978-3-540-85151-6

Vol. 162. Costin Badica, Giuseppe Mangioni, Vincenza Carchiolo and Dumitru Dan Burdescu (Eds.)
Intelligent Distributed Computing, Systems and Applications, 2008
ISBN 978-3-540-85256-8

Vol. 163. Pawel Delimata, Mikhail Ju. Moshkov, Andrzej Skowron and Zbigniew Suraj
Inhibitory Rules in Data Analysis, 2009
ISBN 978-3-540-85637-5

Vol. 164. Nadia Nedjah, Luiza de Macedo Mourelle, Janusz Kacprzyk, Felipe M.G. França and Alberto Ferreira de Souza (Eds.)
Intelligent Text Categorization and Clustering, 2009
ISBN 978-3-540-85643-6

Vol. 165. Djamel A. Zighed, Shusaku Tsumoto, Zbigniew W. Ras and Hakim Hacid (Eds.)
Mining Complex Data, 2009
ISBN 978-3-540-88066-0

Vol. 166. Constantinos Koutsojannis and Spiros Sirmakessis (Eds.)
Tools and Applications with Artificial Intelligence, 2009
ISBN 978-3-540-88068-4

Vol. 167. Ngoc Thanh Nguyen and Lakhmi C. Jain (Eds.)
Intelligent Agents in the Evolution of Web and Applications, 2009
ISBN 978-3-540-88070-7

Vol. 168. Andreas Tolk and Lakhmi C. Jain (Eds.)
Complex Systems in Knowledge-based Environments: Theory, Models and Applications, 2009
ISBN 978-3-540-88074-5

Vol. 169. Nadia Nedjah, Luiza de Macedo Mourelle and Janusz Kacprzyk (Eds.)
Innovative Applications in Data Mining, 2009
ISBN 978-3-540-88044-8

Vol. 170. Lakhmi C. Jain and Ngoc Thanh Nguyen (Eds.)
Knowledge Processing and Decision Making in Agent-Based Systems, 2009
ISBN 978-3-540-88048-6

Vol. 171. Chi-Keong Goh, Yew-Soon Ong and Kay Chen Tan (Eds.)
Multi-Objective Memetic Algorithms, 2009
ISBN 978-3-540-88050-9

Vol. 172. I-Hsien Ting and Hui-Ju Wu (Eds.)
Web Mining Applications in E-Commerce and E-Services, 2009
ISBN 978-3-540-88080-6

Vol. 173. Tobias Grosche
Computational Intelligence in Integrated Airline Scheduling, 2009
ISBN 978-3-540-89886-3

Vol. 174. Ajith Abraham, Rafael Falcón and Rafael Bello (Eds.)
Rough Set Theory: A True Landmark in Data Analysis, 2009
ISBN 978-3-540-89886-3

Vol. 175. Godfrey C. Onwubolu and Donald Davendra (Eds.)
Differential Evolution: A Handbook for Global Permutation-Based Combinatorial Optimization, 2009
ISBN 978-3-540-92150-9

Prof. Godfrey C. Onwubolu
22C Berwick Crescent
Richmond Hill
ON L4C 0B8
Canada

Donald Davendra
Department of Applied Informatics
Tomas Bata Univerzity in Zlin
Nad Stranemi 4511
Zlin 76001
Czech Republic
Email: davendra@fai.utb.cz

ISBN 978-3-540-92150-9   e-ISBN 978-3-540-92151-6

DOI 10.1007/978-3-540-92151-6

Studies in Computational Intelligence   ISSN 1860949X

Library of Congress Control Number: 2008942026

© 2009 Springer-Verlag Berlin Heidelberg

This work is subject to copyright. All rights are reserved, whether the whole or part of the material is concerned, specifically the rights of translation, reprinting, reuse of illustrations, recitation, broadcasting, reproduction on microfilm or in any other way, and storage in data banks. Duplication of this publication or parts thereof is permitted only under the provisions of the German Copyright Law of September 9, 1965, in its current version, and permission for use must always be obtained from Springer. Violations are liable to prosecution under the German Copyright Law.

The use of general descriptive names, registered names, trademarks, etc. in this publication does not imply, even in the absence of a specific statement, that such names are exempt from the relevant protective laws and regulations and therefore free for general use.

Typeset & Cover Design: Scientific Publishing Services Pvt. Ltd., Chennai, India.

Printed in acid-free paper

9 8 7 6 5 4 3 2 1

springer.com

Godfrey C. Onwubolu
Donald Davendra
(Eds.)

Differential Evolution: A Handbook for Global Permutation-Based Combinatorial Optimization

Springer

Godfrey Onwubolu would like to dedicate it to God our Creator, who takes the honor.

Donald Davendra would like to dedicate it to his mother.

Foreword

What is combinatorial optimization? Traditionally, a problem is considered to be combinatorial if its set of feasible solutions is both finite and discrete, i.e., enumerable. For example, the traveling salesman problem asks in what order a salesman should visit the cities in his territory if he wants to minimize his total mileage (see Sect. 2.2.2). The traveling salesman problem's feasible solutions - permutations of city labels - comprise a finite, discrete set. By contrast, Differential Evolution was originally designed to optimize functions defined on real spaces. Unlike combinatorial problems, the set of feasible solutions for real parameter optimization is continuous.

Although Differential Evolution operates internally with floating-point precision, it has been applied with success to many numerical optimization problems that have traditionally been classified as combinatorial because their feasible sets are discrete. For example, the knapsack problem's goal is to pack objects of differing weight and value so that the knapsack's total weight is less than a given maximum and the value of the items inside is maximized (see Sect. 2.2.1). The set of feasible solutions - vectors whose components are nonnegative integers - is both numerical and discrete. To handle such problems while retaining full precision, Differential Evolution copies floating-point solutions to a temporary vector that, prior to being evaluated, is truncated to the nearest feasible solution, e.g., by rounding the temporary parameters to the nearest nonnegative integer.

By truncating real-valued parameters to their nearest feasible value, Differential Evolution can be applied to combinatorial tasks whose parameters are discrete but numerical. If, however, objective function parameters are symbolic - as they are in the traveling salesman problem - then Differential Evolution is inapplicable because it relies in part on arithmetic operators. Unlike the knapsack problem's integral vectors, the permutations that are the traveling salesman problem's feasible solutions cannot be added, subtracted or scaled without first being given a numerical representation ("arithmetized").

In the traveling salesman problem, labels for cities may be drawn from any alphabet and assigned in any order provided that no two cities share the same label. Since cities can be labeled with symbols drawn from any alphabet, they can be distinguished by non-negative integers, in which case feasible solutions for the n-city problem are permutations of the sequence $[0, 1, 2, ..., n-1]$. There is then a sense in which permutations can be added, subtracted and scaled like the knapsack problem's integral vectors. For example, if we assign 0 to Chicago, 1 to Miami, 2 to Denver and 3 to Portland, then the tour Chicago-Miami-Denver-Portland becomes $a = [0,1,2,3]$, while the tour Portland-Denver-Chicago-Miami is $b = [3,2,0,1]$. For this particular mapping of numerals to cities, the modulo 4 difference between the two tours' numerical representations is $(a-b) = [1,3,2,2]$.

Although the freedom to label cities with numerals makes it possible to numerically encode permutations, the freedom to assign numerals in any order means that the "difference" between two permutations depends on the (arbitrary) initial assignment of numerals to cities. If the preceding example had assigned 3 to Chicago, 1 to Miami, 0 to Denver and 2 to Portland, then $a = [3,1,0,2]$, $b = [2,0,3,1]$ and $(a-b)$ modulo 4 becomes $[1,1,1,1]$, not $[1,3,2,2]$. Consequently, this simple numeral assignment scheme fails to provide a meaningful metric because the difference - and by extension, the (Euclidean) distance - between two permutations does not depend on their properties, but on an arbitrary choice. Since Differential Evolution depends heavily on vector/matrix differences, a meaningful metric will be crucial to its success.

Alternatively, permutations can be numerically encoded as n-by-n adjacency matrices (see Sect. 2.3.2). Each city indexes both a row and a column so that the entry in row i and column j is 1 if cities i and j are adjacent in the corresponding permutation and 0 if they are not. Unlike arbitrary numeral assignment, the adjacency matrix numerically encodes a property of the permutation - its connectedness. Furthermore, the modulo 2 difference between two adjacency matrices reveals those connections that the two permutations do not share. If cities are arbitrarily reassigned so that they index different rows and columns, the adjacency matrices for a given pair of permutations will change, but the modulo 2 difference between their new representations will still identify the connections that they do not share. When compared to the nave numeral assignment scheme, the adjacency matrix approach is a more rational way to numerically encode permutations because the difference between two adjacency matrix representations is not arbitrary.

In summary, applying Differential Evolution to numerical combinatorial optimization problems is relatively straightforward. Symbolic combinatorial problems, however, require a property-based function to map symbolic solutions to numerical ones and an inverse transformation to map the numerical solutions proposed by Differential Evolution back into symbolic ones. The configurations generated by these mappings may need "repair" to become feasible, but with well chosen transformations - like those explored in this book - Differential Evolution can be an effective tool for combinatorial optimization.

Vacaville California, September 2008   Kenneth V. Price

Preface

The original classical Differential Evolution (DE) which Storn and Price developed was designed to solve only problems characterized by continuous parameters. This means that only a subset of real-world problems could be solved by the original canonical DE. For quite some time, this deficiency made DE not to be employed to a vast number of real-world problems which characterized by permutative-based combinatorial parameters. Over the years, some researchers have been working in the area of DE permutative-based combinatorial optimization and they have found that DE is quite appropriate for combinatorial optimization and that it is effective and competitive with other approaches in this domain. Some of the DE permutative-based combinatorial optimization approaches that have proved effective include: Forward/Backward Transformation Approach; Relative Position Indexing Approach; Smallest Position Value Approach; Discrete/Binary Approach; and Discrete Set Handling Approach.

To date, there are very few books that present the classical DE which is continuous based, and to the best of our knowledge there is no book that presents DE permutative-based combinatorial optimization approaches. The main purpose of this book therefore is to present the work done by the originators of a number of DE permutative-based combinatorial optimization variants listed above.

The book discusses and differentiates both the continuous space DE formulation and the permutative-based combinatorial DE formulation and shows that these formulations complement each other and none of them is complete on its own. Therefore we have shown that this book complements that of Price et al. (2005) and vice versa. Taken together therefore, both books will be needed by practitioners and students interested in DE in order to have the full potentials of DE at their disposal. In other words, DE as an area of optimization is incomplete unless it can deal with real-life problems in the areas of continuous space as well as permutative-based combinatorial domain.

Chapter authors background: Chapter authors are to the best of our knowledge the originators or closely related to the originators of the above mentioned DE permutative-based combinatorial optimization approaches. Hence, this book will be one of the leading books in DE permutative-based combinatorial optimization approaches.

Organization of the Chapters: Onwubolu and Davendra, present "Motivation for Differential Evolution for Permutative - based Combinatorial Problems" in Chapter 1 as well as "Differential Evolution for Permutation-based Combinatorial Problems", in Chapter 2 in order to show the rationale and give an overview of the book. Onwubolu and Davendra also present "Forward/Backward Transformation Approach" in Chapter 3; Onwubolu is the originator of this approach and Davendra extended its capability to realize the enhanced DE version. Daniel Lichtblau, the originator of "Relative Position Indexing Approach" presents Chapter 4. Tasgetiren, Chen, Gencyilmaz and Gattoufi, the originators of "Smallest Position Value Approach" present Chapter 5. Also, Tasgetiren, Liang, Pan and Suganthan, the originators of "Discrete/Binary Approach" present Chapter 6. Ivan Zelinka the originator of "Discrete Set Handling Approach" presents Chapter 7.

Audience: The book will be an instructional material for senior undergraduate and entry-point graduate students in computer science, applied mathematics, statistics, management and decision sciences, and engineering, who are working in the area of modern optimization. Researchers who want to know how to solve permutative-based combinatorial optimization problems using DE will find this book a very useful handbook and the starting point. The book will be a resource handbook and material for practitioners who want to apply these methods that solve real-life problems to their challenging applications.

Appendix: The book will have all Source Codes for Chapters 3 - 7 in accompanying CD ROM.

Canada and Czech Republic,   Godfrey Onwubolu
September 2008   Donald Davendra

Acknowledgements

We are grateful to the contributors to this book for their willingness to work on this special book project which focuses on non-popular aspect of differential evolution, as well as their ideas and cooperation. From Springer-Verlag in Heidelberg, Germany, we thank Thomas Ditzinger and Heather King for their enthusiasm and editorial guidance throughout the period of our preparing the book.

Godfrey Onwubolu specially thanks his wife Ngozi for bearing with him and supporting him throughout the period of his working on the book project especially during the period of their relocating to a new country at the peak of the book project; their children Chioma, Chineye, Chukujindu, Chinwe, and Chinedu are all appreciated for bearing with him as usual.

Donald Davendra would like to acknowledge the following people for their help and support. Firstly, to his parents, Michael Davendra and Manjula Devi, the two pillars of his life, for believing in him and letting him follow his dreams, and to his sister, Annjelyn Shalvina, for her love and support when it mattered most. Secondly, to his mentors, Prof Godfrey Onwubolu, who first led him down the path of Differential Evolution, and Prof Ivan Zelinka, who took a leap of faith in him, and showed him the humorous side of research. Na końcu chciałby podziękować Mgr. Magdalenie Bialic, która zawsze wspiera go swoja miłościa i dzieki której jest szcześliwy. He would like to dedicate all his contribution in this book to his mother.

Godfrey Onwubolu and Donald Davendra acknowledge Rainer Storn for his supportive comments on the book project during initial communication with him.

Contents

1 Motivation for Differential Evolution for Permutative–Based Combinatorial Problems 1
Godfrey Onwubolu, Donald Davendra
  1.1 Introduction 1
    1.1.1 Continuous Space Optimization DE Problems 2
    1.1.2 Permutative–Based Combinatorial Optimization DE Problem 2
    1.1.3 Suitability of Differential Evolution as a Combinatorial Optimizer 3
  1.2 Canonical Differential Evolution for Continuous Optimization Problems 4
  1.3 Differential Evolution for Permutative–Based Combinatorial Optimization Problems 9
  1.4 Conclusions 10
  References 11

2 Differential Evolution for Permutation–Based Combinatorial Problems 13
Godfrey Onwubolu, Donald Davendra
  2.1 Introduction 13
    2.1.1 Wide-Sense Combinatorial Optimization 13
    2.1.2 Strict-Sense Combinatorial Optimization 14
    2.1.3 Feasible Solutions versus "Repairing" Infeasible Solutions for Strict-Sense Combinatorial Optimization 14
  2.2 Combinatorial Problems 14
    2.2.1 Knapsack Problem 15
    2.2.2 Travelling Salesman Problem (TSP) 15
    2.2.3 Automated Drilling Location and Hit Sequencing 20
    2.2.4 Dynamic Pick and Place (DPP) Model of Placement Sequence and Magazine Assignment 22

		2.2.5	Vehicle Routing Problem	25
		2.2.6	Facility Location Problem	25
	2.3	Permutation-Based Combinatorial Approaches		26
		2.3.1	The Permutation Matrix Approach	26
		2.3.2	Adjacency Matrix Approach	27
		2.3.3	Relative Position Indexing	27
		2.3.4	Forward/Backward Transformation Approach	28
		2.3.5	Smallest Position Value Approach	29
		2.3.6	Discrete/Binary Approach	30
		2.3.7	Discrete Set Handling Approach	31
		2.3.8	Anatomy of Some Approaches	31
	2.4	Conclusions		32
	References			33
3	**Forward Backward Transformation**			**35**
	Donald Davendra, Godfrey Onwubolu			
	3.1	Introduction		35
	3.2	Differential Evolution		36
		3.2.1	Tuning Parameters	38
	3.3	Discrete Differential Evolution		38
		3.3.1	Permutative Population	39
		3.3.2	Forward Transformation	39
		3.3.3	Backward Transformation	40
		3.3.4	Recursive Mutation	40
	3.4	Enhanced Differential Evolution		41
		3.4.1	Repairment	42
		3.4.2	Improvement Strategies	45
		3.4.3	Local Search	46
	3.5	Worked Example		48
	3.6	Flow Shop Scheduling		59
		3.6.1	Flow Shop Scheduling Example	60
		3.6.2	Experimentation for Discrete Differential Evolution Algorithm	62
		3.6.3	Experimentation for Enhanced Differential Evolution Algorithm	65
	3.7	Quadratic Assignment Problem		68
		3.7.1	Quadratic Assignment Problem Example	69
		3.7.2	Experimentation for Irregular QAP	71
		3.7.3	Experimentation for Regular QAP	72
	3.8	Traveling Salesman Problem		73
		3.8.1	Traveling Salesman Problem Example	74
		3.8.2	Experimentation on Symmetric TSP	76
		3.8.3	Experimentation on Asymmetric TSP	76
	3.9	Analysis and Conclusion		77
	References			78

4 Relative Position Indexing Approach 81
Daniel Lichtblau
- 4.1 Introduction 81
- 4.2 Two Simple Examples 84
  - 4.2.1 Pythagorean Triples 84
- 4.3 Maximal Determinants 86
- 4.4 Partitioning a Set 88
  - 4.4.1 Set Partitioning via Relative Position Indexing 90
  - 4.4.2 Set Partitioning via Knapsack Approach 93
  - 4.4.3 Discussion of the Two Methods 95
- 4.5 Minimal Covering of a Set by Subsets 95
  - 4.5.1 An Ad Hoc Approach to Subset Covering 96
  - 4.5.2 Subset Covering via Knapsack Formulation 98
- 4.6 An Assignment Problem 101
  - 4.6.1 Relative Position Indexing for Permutations 104
  - 4.6.2 Representing and Using Permutations as Shuffles 106
  - 4.6.3 Another Shuffle Method 109
- 4.7 Hybridizing Differential Evolution for the Assignment Problem 112
- 4.8 Future Directions 118
- References 119

5 Smallest Position Value Approach 121
Fatih Tasgetiren, Angela Chen, Gunes Gencyilmaz, Said Gattoufi
- 5.1 Introduction 121
- 5.2 Differential Evolution Algorithm 123
  - 5.2.1 Solution Representation 125
  - 5.2.2 An Example Instance of the GTSP 126
  - 5.2.3 Complete Computational Procedure of DE 127
- 5.3 Insertion Methods 129
  - 5.3.1 Hybridization with Local Search 131
- 5.4 Computational Results 132
- 5.5 Conclusions 136
- References 137

6 Discrete/Binary Approach 139
Fatih Tasgetiren, Yun-Chia Liang, Quan-Ke Pan, Ponnuthurai Suganthan
- 6.1 Introduction 139
- 6.2 Discrete Differential Evolution Algorithm 141
  - 6.2.1 Solution Representation 144
  - 6.2.2 Complete Computational Procedure of DDE 145
  - 6.2.3 NEH Heuristic 146
  - 6.2.4 Insertion Methods 147
  - 6.2.5 Destruction and Construction Procedure 149
  - 6.2.6 PTL Crossover Operator 151

		6.2.7	Insert Mutation Operator	151
		6.2.8	DDE Update Operations	152
	6.3	Hybridization with Local Search		153
	6.4	Computational Results		154
		6.4.1	Solution Quality	155
		6.4.2	Computation Time	157
		6.4.3	Comparison to Other Algorithms	157
	6.5	Conclusions		160
	References			160
7	**Discrete Set Handling**			163
	Ivan Zelinka			
	7.1	Introduction		163
	7.2	Permutative Optimization		164
		7.2.1	Travelling Salesman Problem	164
		7.2.2	Flow Shop Scheduling Problem	165
		7.2.3	2 Opt Local Search	165
	7.3	Discrete Set Handling and Its Application		166
		7.3.1	Introduction and Principle	166
		7.3.2	DSH Applications on Standard Evolutionary Algorithms	167
		7.3.3	DSH Applications on Class of Genetic Programming Techniques	169
	7.4	Differential Evolution in Mathematica Code		174
		7.4.1	DE Flow Shop Scheduling	182
		7.4.2	DE Traveling Salesman Problem	183
	7.5	DE Example		184
		7.5.1	Initialization	185
		7.5.2	DSH Conversion	185
		7.5.3	Fitness Evaluation	186
		7.5.4	DE Application	186
	7.6	Experimentation		189
		7.6.1	Flow Shop Scheduling Tuning	190
		7.6.2	Traveling Salesman Problem Tuning	194
		7.6.3	Flow Shop Scheduling Results	197
		7.6.4	Traveling Salesman Problem Results	201
	7.7	Conclusion		203
	References			203
A	**Smallest Position Value Approach**			207
	A.1	Clusters for the Instance 11EIL51		207
	A.2	Pseudo Code for Distance Calculation		207
	A.3	Distance (ij, d_{ij}) Information for the Instance 11EIL51		208
Author Index				213

List of Contributors

Angela Chen
Department of Finance, Nanya Institute of Technology,
Taiwan 320, R.O.C,
achen@nanya.edu.tw

Donald Davendra
Faculty of Applied Informatics, Tomas Bata Univerzity in Zlin, Nad Stranemi 4511, Zlin 76001, Czech Republic,
davendra@fai.utb.cz

Gunes Gencyilmaz
Department of Management, Istanbul Kultur University, Istanbul, Turkey,
g.gencyilmaz@iku.edu.tr

Said Gattoufi
Department of Operations Management and Business Statistics, Sultan Qaboos University,
Muscat, Sultanate of Oman,
gattoufi@squ.edu.om

Yun-Chia Liang
Department of Industrial Engineering and Management, Yuan Ze University,
Taiwan, R.O.C,
ycliang@saturn.yzu.edu.tw

Daniel Lichtblau
Wolfram Research, Inc., 100 Trade Center Dr., Champaign,
IL 61820, USA,
danl@wolfram.com

Godfrey Onwubolu Knowledge Management & Mining,
Inc., Richmond Hill, Ontario, Canada,
onwubolu_g@dsgm.ca

Quan-Ke Pan
College of Computer Science,
Liaocheng University, Liaocheng,
P. R. China,
qkpan@lctu.edu.cn

Ponnuthurai Suganthan
School of Electrical and Electronic Engineering, Nanyang Technological University, Singapore,
epnsugan@ntu.edu.sg

FatihTasgetiren
Department of Operations Management and Business Statistics, Sultan Qaboos University,
Muscat, Sultanate of Oman,
mfatih@squ.edu.om

Ivan Zelinka
Faculty of Applied Informatics, Tomas Bata Univerzity in Zlin, Nad Stranemi 4511, Zlin 76001, Czech Republic,
zelinka@fai.utb.cz

1

Motivation for Differential Evolution for Permutative−Based Combinatorial Problems

Godfrey Onwubolu[1] and Donald Davendra[2]

[1] Knowledge Management & Mining, Inc., Richmond Hill, Ontario, Canada
    `onwubolu_g@dsgm.ca`
[2] Tomas Bata Univerzity in Zlin, Faculty of Applied Informatics, Nad Stranemi 4511,
    Zlin 76001, Czech Republic
    `davendra@fai.utb.cz`

Abstract. It is generally accepted that Differential Evolution (DE) was originally designed to solve problems which are defined in continuous form. Some researchers have however, felt that this is a limiting factor on DE, hence there have been vigorous research work to extend the functionalities of DE to include permutative-based combinatorial problems. This chapter sets the scene for the book by discussing the motivation for presenting the foundational theories for a number of variants of DE for permutative-based combinatorial problems. These DE variants are presented by their initiators or proposers, to the best of our knowledge.

1.1 Introduction

Whether in industry or in research, users generally demand that a practical optimization technique should fulfil three requirements:

1. the method should find the true global minimum, regardless of the initial system parameter values;
2. convergence should be fast; and
3. the program should have a minimum of control parameters so that it will be easy to use.

[2] invented the differential evolution (DE) algorithm in a search for a technique that would meet the above criteria. DE is a method, which is not only astonishingly simple, but also performs extremely well on a wide variety of test problems. It is inherently parallel because it is a population based approach and hence lends itself to computation via a network of computers or processors. The basic strategy employs the difference of two randomly selected parameter vectors as the source of random variations for a third parameter vector.

There are broadly speaking two types of real−world problems that may be solved:

1. those that are characterized by continuous parameters; and
2. those that are characterized by permutative-based combinatorial parameters..

This classification is important in the context of the motivation for this book. The original canonical DE which Storn and Price developed was designed to solve only

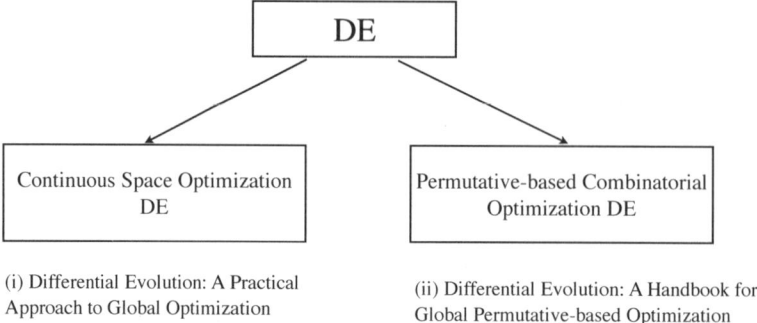

Fig. 1.1. DE framework (i) Existing book (Price et al. 2005); (ii) Present book

problems characterized by continuous parameters. This means that only a subset of real-world problems could be solved by the original canonical DE. For quite some time, this deficiency made DE not to be employed to a vast number of real-world problems which characterized by permutative-based combinatorial parameters. Fig 1.1 shows the framework into which the current book fits, showing that there are two mainstreams of philosophical schools that need to be considered in presenting DE for solving real-world problems. This framework is important as it shows that the current book compliments the first book on DE which only addresses one aspect: continuous parameters.

1.1.1 Continuous Space Optimization DE Problems

A typical continuous space optimization DE problem is the generalized Rosenbrock function given as:

$$f(x) = \sum_{j=0}^{D-2} \left(100. \left(x_{j+1} - x_j^2\right)^2 + (x_j - 1)^2 \right),$$
$$-30 \leq x \leq 30, \quad j = 0, 1, ..., D-1, \quad D > 1,$$
$$f(x^*) = 0, \quad x_j^* = 1, \quad \varepsilon = 1.0 \times 10^{-6}. \tag{1.1}$$

The solution of this problem is shown in Fig 1.2.

1.1.2 Permutative–Based Combinatorial Optimization DE Problem

A typical permutative-based combinatorial optimization DE problem is the flow shop scheduling five-job-four machine problem whose operation times are shown in Table 1.1 The objective is to find the best sequence to realize the optimal completion time (makespan). As we see, this problem is very different from the continuous space problem because we are interested in sequence such as 1, 2, 3, 4, 5 which is permutative in nature.

The minimization of completion time (makespan) for a flow shop schedule is equivalent to minimizing the objective function $\Im$:

$$\Im = \sum_{j=1}^{n} C_{m,j} \tag{1.2}$$

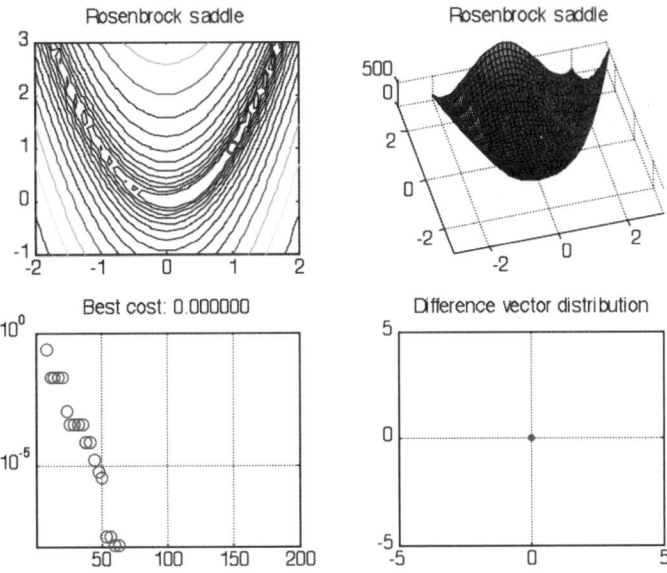

Fig. 1.2. The generalized Rosenbrock function problem

Table 1.1. Example of job times

jobs	j_1	j_2	j_3	j_4	j_5
P_{1,j_k}	6	4	4	5	1
P_{2,j_k}	4	6	2	4	3
P_{3,j_k}	3	3	4	1	3
P_{4,j_k}	4	4	5	3	1

such that:
$$C_{i,j} = \max\left(C_{i-1,j},\, C_{i,j-1}\right) + P_{i,j} \qquad (1.3)$$

For this problem the schedule of $\{2,1,4,3,5\}$ gives a fitness (makespan) of 31. The solution includes a permutative schedule as well as the fitness. It is not feasible to solve this kind of problem using the canonical DE that solves only continuous problems as already discussed.

1.1.3 Suitability of Differential Evolution as a Combinatorial Optimizer

An extract from the summary of Section 4.4 of [2] reads:

> "Although DE has performed well on wide-sense combinatorial problems, its suitability as a combinatorial optimizer is still a topic of considerable debate and a definitive judgment cannot be given at this time".

Over the years, some researchers have been working in the area of DE permutative–based combinatorial optimization and they have found that DE is quite appropriate for combinatorial optimization and that it is effective and competitive with other approaches in this domain. Some of the DE permutative–based combinatorial optimization approaches that have proved effective include:

1. Forward/Backward Transformation Approach;
2. Relative Position Indexing Approach;
3. Smallest Position Value Approach;
4. Discrete/Binary Approach; and
5. Discrete Set Handling Approach.

These approaches have been applied to combinatorial optimization problems with competitive results when compared to other optimization approaches, and they form the basis for writing this book. The remainder of this book explores available DE approaches for solving permutative-based combinatorial problems. Although there have been discussions regarding DE approaches that rely to varying degrees on repair mechanisms, it is now generally agreed that in order to solve permutative-based combinatorial problems, it is necessary to employ some heuristics as some other evolutionary computation approaches do, rather than insisting on approaches that generate only feasible solutions. Each method proposes an analog of DE's differential mutation operator to solve permutative-based combinatorial problems.

1.2 Canonical Differential Evolution for Continuous Optimization Problems

The parameters used in DE are $\Im$ = cost or the value of the objective function, D = problem dimension, NP = population size, P = population of X–vectors, G = generation number, $Gmax$ = maximum generation number, X = vector composed of D parameters, V = trial vector composed of D parameters, CR = crossover factor. Others are F = scaling factor ($0 < F \leq 1.2$), (U) = upper bound, (L) = lower bound, **u**, and **v** = trial vectors, $x_{best}^{(G)}$ = vector with minimum cost in generation G, $x_i^{(G)}$ = ith vector in generation G, $b_i^{(G)}$ = ith buffer vector in generation G, $x_{r1}^{(G)}, x_{r2}^{(G)}$ = randomly selected vector, L = random integer ($0 < L < D-1$). In the formulation, N = number of cities. Some integers used are i, j.

Differential Evolution (DE) is a novel parallel direct search method, which utilizes NP parameter vectors

$$x_i^{(G)}, i = 0, 1, 2, ..., NP-1 \qquad (1.4)$$

as a population for each generation, G. The population size, NP does not change during the minimization process. The initial population is generated randomly assuming a uniform probability distribution for all random decisions if there is no initial intelligent information for the system. The crucial idea behind DE is a new scheme for generating trial parameter vectors. DE generates new parameter vectors by adding the weighted difference vector between two population members to a third member. If the resulting vector yields a lower objective function value than a predetermined population member, the newly generated vector replaces the vector with which it was compared. The comparison vector can, but need not be part of the generation process mentioned above. In addition the best parameter vector $x_{best}^{(G)}$, is evaluated for every generation G in order to keep track of the progress that is made during the minimization process. Extracting distance and direction information from the population to generate random deviations result in an adaptive scheme with excellent convergence properties [3].

Descriptions for the earlier two most promising variants of DE (later known as DE2 and DE3) are presented in order to clarify how the search technique works, then a complete list of the variants to date are given thereafter. The most comprehensive book that describes DE for continuous optimization problems is [2].

Scheme DE2

Initialization
As with all evolutionary optimization algorithms, DE works with a population of solutions, not with a single solution for the optimization problem. Population P of generation G contains NP solution vectors called individuals of the population and each vector represents potential solution for the optimization problem:

$$P^{(G)} = X_i^{(G)} \quad i = 1,...,NP; \; G = 1,...,G_{\max} \tag{1.5}$$

Additionally, each vector contains D parameters:

$$X_i^{(G)} = x_{j,i}^{(G)} \quad i = 1,...,NP; \; j = 1,...,D \tag{1.6}$$

In order to establish a starting point for optimum seeking, the population must be initialized. Often there is no more knowledge available about the location of a global optimum than the boundaries of the problem variables. In this case, a natural way to initialize the population $P^{(0)}$ (initial population) is to seed it with random values within the given boundary constraints:

$$P^{(0)} = x_{j,i}^{(0)} = x_j^{(L)} + rand_j[0,1] \bullet \left(x_j^{(U)} - x_j^{(L)} \right) \; \forall i \in [1,NP]; \forall j \in [1,D] \tag{1.7}$$

where $rand_j[0,1]$ represents a uniformly distributed random value that ranges from zero to one. The lower and upper boundary constraints are, $X^{(L)}$ and $X^{(L)}$, respectively:

$$x_j^{(L)} \leq x_j \leq x_j^{(U)} \; \forall j \in [1,D] \tag{1.8}$$

For this scheme and other schemes, three operators are crucial: mutation, crossover and selection. These are now briefly discussed.

Mutation

The first variant of DE works as follows: for each vector $x_i^{(G)}, i = 0,1,2,..,NP-1$, a trial vector v is generated according to

$$v_{j,i}^{(G+1)} = x_{j,r1}^{(G)} + F \bullet \left(x_{j,r2}^{(G)} - x_{j,r3}^{(G)} \right) \tag{1.9}$$

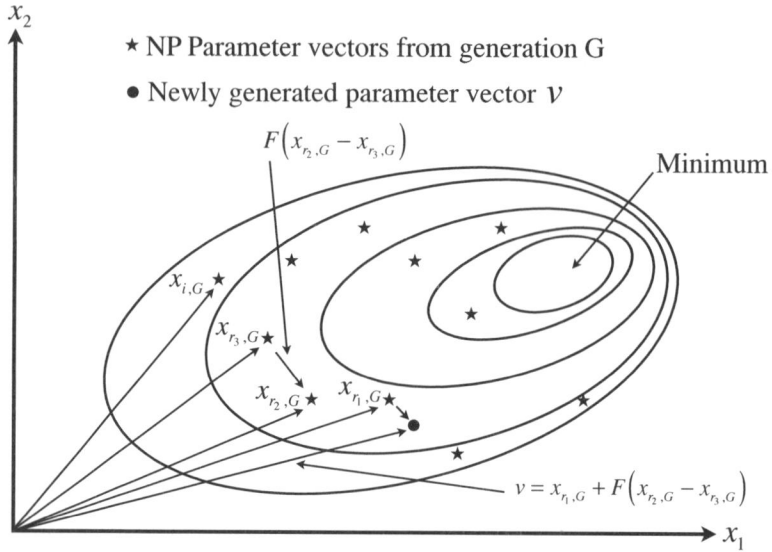

Fig. 1.3. Contour lines and the process for generating v in scheme DE1

where $i \in [1, NP]$; $j \in [1, D]$, $F > 0$, and the integers $r1$, $r2$, $r3 \in [1, NP]$ are generated randomly selected, except: $r1 \neq r2 \neq r3 \neq i$.

Three randomly chosen indexes, $r1$, $r2$, and $r3$ refer to three randomly chosen vectors of population. They are mutually different from each other and also different from the running index i. New random values for $r1$, $r2$, and $r3$ are assigned for each value of index i (for each vector). A new value for the random number $rand[0, 1]$ is assigned for each value of index j (for each vector parameter). F is a real and constant factor, which controls the amplification of the differential variation. A two dimensional example that illustrates the different vectors which play a part in DE2 are shown in Fig 1.3.

Crossover

In order to increase the diversity of the parameter vectors, the vector

$$u = (u_1, u_2, ..., u_D)^T \quad (1.10)$$

$$u_j^{(G)} = \begin{cases} v_j^{(G)} & \text{for } j = \langle n \rangle_D, \langle n+1 \rangle_D, ..., \langle n+L-1 \rangle_D \\ \left(x_i^{(G)}\right)_j & \text{otherwise} \end{cases} \quad (1.11)$$

is formed where the acute brackets $\langle \rangle_D$ denote the modulo function with modulus D. This means that a certain sequence of the vector elements of u are identical to the elements of v, the other elements of u acquire the original values of $x_i^{(G)}$. Choosing a subgroup of parameters for mutation is similar to a process known as crossover in genetic algorithm. This idea is illustrated in Fig 1.4 for D = 7, n = 2 and L = 3. The starting index n in (12)

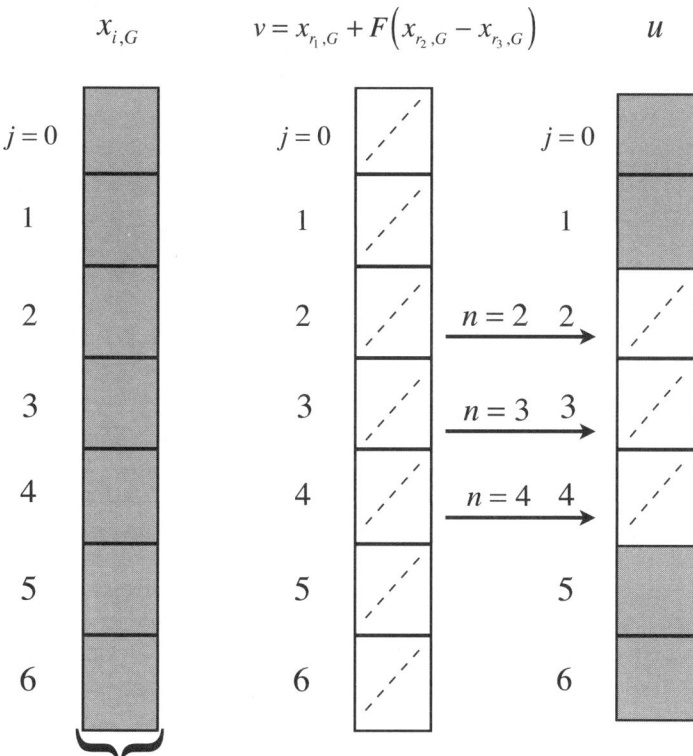

Parameter vector containing the parameters
$x_j, j = \{0, 1, ..., D-1\}$

Fig. 1.4. Crossover process for D = 7, n = 2 and L = 3

is a randomly chosen integer from the interval [0, D-1]. The integer L is drawn from the interval [0, D-1] with the probability $\Pr(L = v) = (CR)^v$. $CR \in [0, 1]$ is the crossover probability and constitutes a control variable for the DE2-scheme. The random decisions for both n and L are made anew for each trial vector v.

Crossover

In order to decide whether the new vector u shall become a population member of generation $G+1$, it will be compared to $x_i^{(G)}$. If vector u yields a smaller objective function value than $x_i^{(G)}$, $x_i^{(G+1)}$ is set to u, otherwise the old value $x_i^{(G)}$ is retained.

Scheme DE3

Basically, scheme DE3 works the same way as DE2 but generates the vector v according to

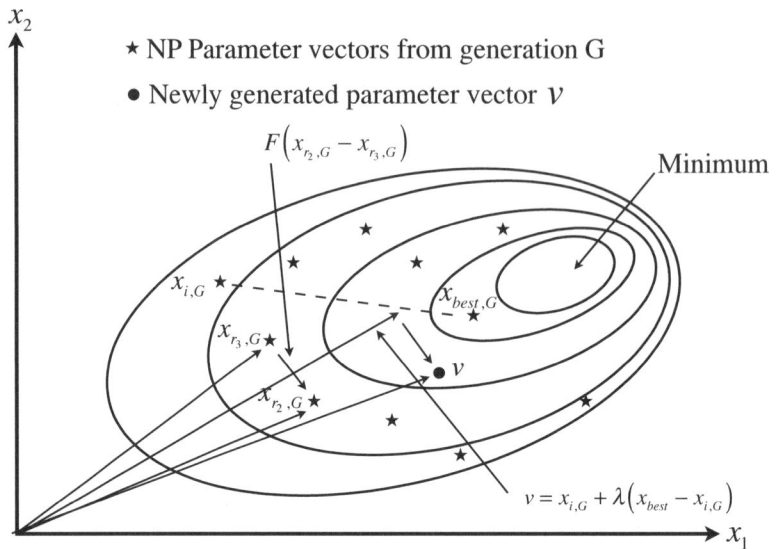

Fig. 1.5. Contour lines and the process for generating v in scheme DE3.

$$v = x_i^{(G)} + \lambda \bullet \left(x_{best}^{(G)} - x_i^{(G)}\right) + F \bullet \left(x_{r2}^{(G)} - x_{r3}^{(G)}\right) \quad (1.12)$$

introducing an additional control variable λ. The idea behind λ is to provide a means to enhance the greediness of the scheme by incorporating the current best vector $x_{best}^{(G)}$. This feature can be useful for non–critical objective functions. Fig 1.5 illustrates the vector–generation process defined by Equation 1.12. The construction of u from v and as well as the decision process are identical to DE2.

DE Strategies

[1] have suggested ten different working strategies of DE and some guidelines in applying these strategies for any given problem (see Table1.2). Different strategies can be adopted in the DE algorithm depending upon the type of problem for which it is applied. The strategies can vary based on the vector to be perturbed, number of difference vectors considered for perturbation, and finally the type of crossover used.

The general convention used above is as follows: DE/x/y/z. DE stands for differential evolution algorithm, x represents a string denoting the vector to be perturbed, y is the number of difference vectors considered for perturbation of x, and z is the type of crossover being used. Other notations are exp: exponential; bin: binomial). Thus, the working algorithm outline by [2] is the seventh strategy of DE, that is, DE/rand/1/bin. Hence the perturbation can be either in the best vector of the previous generation or in any randomly chosen vector. Similarly for perturbation, either single or two vector differences can be used. For perturbation with a single vector difference, out of the three distinct randomly chosen vectors, the weighted vector differential of any two vectors

Table 1.2. DE Strategies

Strategy	Formulation
Strategy 1: DE/best/1/exp:	$v = x_{best}^{(G)} + F \bullet \left(x_{r2}^{(G)} - x_{r3}^{(G)} \right)$
Strategy 2: DE/rand/1/exp:	$v = x_{r1}^{(G)} + F \bullet \left(x_{r2}^{(G)} - x_{r3}^{(G)} \right)$
Strategy 3: DE/rand-to-best/1/exp	$v = x_{i}^{(G)} + \lambda \bullet \left(x_{best}^{(G)} - x_{i}^{(G)} \right) + F \bullet \left(x_{r1}^{(G)} - x_{r2}^{(G)} \right)$
Strategy 4: DE/best/2/exp:	$v = x_{best}^{(G)} + F \bullet \left(x_{r1}^{(G)} + x_{r2}^{(G)} - x_{r3}^{(G)} - x_{r4}^{(G)} \right)$
Strategy 5: DE/rand/2/exp:	$v = x_{r5}^{(G)} + F \bullet \left(x_{r1}^{(G)} + x_{r2}^{(G)} - x_{r3}^{(G)} - x_{r4}^{(G)} \right)$
Strategy 6: DE/best/1/bin:	$v = x_{best}^{(G)} + F \bullet \left(x_{r2}^{(G)} - x_{r3}^{(G)} \right)$
Strategy 7: DE/rand/1/bin:	$v = x_{r1}^{(G)} + F \bullet \left(x_{r2}^{(G)} - x_{r3}^{(G)} \right)$
Strategy 8: DE/rand-to-best/1/bin:	$v = x_{i}^{(G)} + \lambda \bullet \left(x_{best}^{(G)} - x_{i}^{(G)} \right) + F \bullet \left(x_{r1}^{(G)} - x_{r2}^{(G)} \right)$
Strategy 9: DE/best/2/bin	$v = x_{best}^{(G)} + F \bullet \left(x_{r1}^{(G)} + x_{r2}^{(G)} - x_{r3}^{(G)} - x_{r4}^{(G)} \right)$
Strategy 10: DE/rand/2/bin:	$v = x_{r5}^{(G)} + F \bullet \left(x_{r1}^{(G)} + x_{r2}^{(G)} - x_{r3}^{(G)} - x_{r4}^{(G)} \right)$

is added to the third one. Similarly for perturbation with two vector differences, five distinct vectors other than the target vector are chosen randomly from the current population. Out of these, the weighted vector difference of each pair of any four vectors is added to the fifth one for perturbation.

In exponential crossover, the crossover is performed on the D (the dimension or number of variables to be optimized) variables in one loop until it is within the CR bound. For discrete optimization problems, the first time a randomly picked number between 0 and 1 goes beyond the CR value, no crossover is performed and the remaining D variables are left intact. In binomial crossover, the crossover is performed on each the D variables whenever a randomly picked number between 0 and 1 is within the CR value. Hence, the exponential and binomial crossovers yield similar results.

1.3 Differential Evolution for Permutative–Based Combinatorial Optimization Problems

The canonical DE cannot be applied to discrete or permutative problems without modification. The internal crossover and mutation mechanism invariably change any applied value to a real number. This in itself will lead to infeasible solutions. The objective then becomes one of transformation, either that of the population or that of the internal crossover/mutation mechanism of DE. A number of researchers have decided not to modify in any way the operation of DE strategies, but to manipulate the population in such a way as to enable DE to operate unhindered. Since the solution for a population is permutative, suitable conversion routines are required in order to change the solution from integer to real and then back to integer after crossover.

Application areas where DE for permutative-based combinatorial optimization problems can be applied include but not limited to the following:

Table 1.3. Building blocks and the enhanced versions of DE

Building Blocks	Enhanced DE versions	Chapter
Forward/Backward Transformation Approach	Enhanced DE (EDE)	3
Relative Position Indexing Approach	HPS	4
Smallest Position Value Approach	-	5
Discrete/Binary Approach	-	6
Discrete Set Handling Approach	DE DSH	7

1. Scheduling: Flow Shop, Job Shop, etc.
2. Knapsack
3. Linear Assignment Problem (LAP)
4. Quadratic Assignment Problem (QAP)
5. Traveling Salesman Problem (TSP)
6. Vehicle Routine Problem (VRP)
7. Dynamic pick-and-place model of placement sequence and magazine assignment in robots

In this book, some methods for realizing DE for permutative-based combinatorial optimization problems that are presented in succeeding chapters are as follows:

1. Forward/Backward Transformation Approach: [chapter 3];
2. Relative Position Indexing Approach: [chapter 4];
3. Smallest Position Value Approach: [chapter 5];
4. Discrete/Binary Approach: [chapter 6]; and
5. Discrete Set Handling Approach: [chapter 7].

While the above−listed foundations have been presented in the book, it should be mentioned that a number of enhancement routines have been realized which are based on these fundamental building blocks. For example, the enhanced DE (EDE) is based on fundamentals of the forward/backward transformation approach presented in chapter 3. This philosophy threads throughout the book and should be borne in mind when reading the chapters. Consequently, we have the building blocks and the enhanced versions of DE listed in Table 1.3.

1.4 Conclusions

This chapter has discussed and differentiated both the continuous space DE formulation and the permutative-based combinatorial DE formulation and shown that these formulations compliment each other and none of them is complete on its own. Therefore we have shown that this book complements that of [2] and vice versa. Taken together therefore, both books will be needed by practitioners and students interested in DE in order to have the full potentials of DE at their disposal. In other words, DE as an area of optimization is incomplete unless it can deal with real-life problems in the areas of continuous space as well as permutative-based combinatorial domain.

At least five DE permutative-based combinatorial optimization approaches that have proved effective have been presented in this book and they have been used to solve real-life problems. The results obtained are found to be quite competitive for each approach presented in chapters 3 to 7. Some of these approaches have become the building blocks for realizing higher-order or enhanced version of DE permutative-based combinatorial optimization approaches. Some of these enhanced versions have presented in some of the chapters. Their results are better in terms of quality than the basic versions; and some cases computation times have been drastically reduced when these enhanced version are used for solving the same real-life problems which the basic versions are used for.

References

1. Price, K., Storn, R.: Differential evolution homepage (2001) (Cited September 10, 2008),
 http://www.ICSI.Berkeley.edu/~storn/code.html
2. Price, K., Storn, R., Lampinen, J.: Differential Evolution. Springer, Heidelberg (2005)
3. Storn, R., Price, K.: Differential evolution – a simple and efficient adaptive scheme for global optimization over continuous spaces, Technical Report TR-95-012, ICSI March 1999 (1995) (available via ftp),
 ftp.icsi.berkeley.edu/pub/techreports/1995/tr-95-012.ps.Z

2

Differential Evolution for Permutation–Based Combinatorial Problems

Godfrey Onwubolu[1] and Donald Davendra[2]

[1] Knowledge Management & Mining, Inc., Richmond Hill, Ontario, Canada
  onwubolu_g@dsgm.ca
[2] Tomas Bata Univerzity in Zlin, Faculty of Applied Informatics, Nad Stranemi 4511,
  Zlin 76001, Czech Republic
  davendra@fai.utb.cz

Abstract. The chapter clarifies the differences between wide-sense combinatorial optimization and strict-sense combinatorial optimization and then presents a number of combinatorial problems encountered in practice. Then overviews of the different permutative-based combinatorial approaches presented in the book are given. The chapter also includes an anatomy of the different permutative-based combinatorial approaches in the book, previously carried out elsewhere to show their strengths and weaknesses.

2.1 Introduction

It is first necessary to define what combinatorial problems are. In combinatorial problems, parameters can assume only a finite number of discrete states, so the number of possible vectors is also finite. Several classic algorithmic problems of a purely combinatorial nature include sorting and permutation generation, both of which were among the first non–numerical problems arising on electronic computers. A permutation describes an arrangement, or ordering, of parameters that define a problem. Many algorithmic problems tend to seek the best way to order a set of objects. Any algorithm for solving such problems exactly must construct a series of permutations. [10] classify wide-sense and strict–sense combinatorial optimization.

2.1.1 Wide-Sense Combinatorial Optimization

Consider switching networks which can be divided into fully connected non-blocking networks, and fully connected but blocking networks. Non–blocking switching networks can be re–arrangeable non-blocking networks, wide-sense non-blocking networks, and strictly non-blocking networks. A network is classified as rearrangeable if any idle input may be connected to any idle output provided that existing connections are allowed to be rearranged. A strictly non–blocking network on the other hand is always able to connect any idle input to any idle output without interfering with the existing connections. Wide–sense non-blocking network achieves strictly non-blocking property with the help of an algorithm.

Let us consider another example. In this case we consider nuts having pitch diameters which are expressed in decimal places such as 2.5 mm, 3.6 mm,..., 5.3 mm, 6.2 mm, etc. These nuts are grouped into classes so that Class A nuts belong to those nuts whose pitch diameters lie between 2.5 mm−3.6 mm, Class B nuts belong to those nuts whose pitch diameters lie between 3.65 mm−4.8 mm, etc. In this case the classes are wide sense but the actual dimensions are continuous in nature. Therefore picking nuts based on their classes could be viewed as a wide−sense combinatorial problem because dimensional properties of a nut are continuous variables.

2.1.2 Strict-Sense Combinatorial Optimization

There are a number of **strict-sense** combinatorial problems, such as the traveling salesman problem, the knapsack problem, the shortest-path problem, facility layout problem, vehicle routing problem, etc. These are **strict-sense** combinatorial problems because they have no continuous counterpart [10]. These **strict-sense** combinatorial problems require some permutation of some sort. The way the objects are arranged may affect the overall performance of the system being considered. Arranging the objects incorrectly may affect the overall performance of the system. If there is a very large number of the object, then the number of ways of arranging the objects introduces another dimension of problem known as combinatorial explosion. Classical DE was not designed to solve this type of problem because these problems have hard constraints. Strong constraints like those imposed in the traveling salesman problem or facility layout problem make **strict-sense** combinatorial problems notoriously difficult for any optimization algorithm. It is this class of problems that this book is aimed at solving. A number of techniques have been devised to stretch the capabilities of DE to solve this type of hard constraint-type problems.

2.1.3 Feasible Solutions versus "Repairing" Infeasible Solutions for Strict-Sense Combinatorial Optimization

In DE's case, the high proportion of infeasible vectors caused by constraints prevents the population from thoroughly exploring the objective function surface. [10] concluded that in order to minimize the problems posed by infeasible vectors, algorithms can either generate only feasible solutions, or "repair" infeasible ones.

The opinion expressed in this book is that all good heuristics are able to transform a combinatorial problem into a space which is amenable for search, and that there is no such thing as an "all-cure" algorithm for combinatorial problems. For example, particle swarm optimization (PSO) works fairly well for combinatorial problems, but only in combination with a good tailored heuristic (see for example, [8]). If such a heuristic is used, then PSO can locate promising regions. The same logic applies to a number of optimization approaches.

2.2 Combinatorial Problems

A wide range of strict-sense combinatorial problems exist for which the classical DE approach cannot solve because these problems are notoriously difficult for any optimization algorithm. In this section, some of these problems are explained and their objective functions are formulated. The knapsack problem, travelling salesman problem (TSP), drilling location and hit sequencing, dynamic pick and place (DPP) model in robotic environment, vehicle routing problem (VRP), and facility layout problem, which are examples of strict-sense combinatorial problems, are discussed in this sub-section.

2.2.1 Knapsack Problem

For example, the single constraint (bounded) knapsack problem reflects the dilemma faced by a hiker who wants to pack as many valuable items in his or her knapsack as possible without exceeding the maximum weight he or she can carry. In the knapsack problem, each item has a weight, w_j, and a value, c_j (Equation 2.1); the constraints are in (Equation 2.1). The goal is to maximize the value of items packed without exceeding the maximum weight, b. The term represents the number of items with weight w_j and value, c_j:

maximize:
$$\sum_{j=0}^{D-1} c_j x_j, \quad x_j \geq 0, \text{integers} \tag{2.1}$$

subject to:
$$\sum_{j=0}^{D-1} w_j x_j \leq b, \quad w_j \geq 0, \ b > 0. \tag{2.2}$$

The solution to this problem will be a set of integers that indicate *how many* items of each type should be packed. As such, the knapsack problem is a strict-sense combinatorial problem because its parameters are discrete, solutions are constrained and it has no continuous counterpart (only a whole number of items can be placed in the knapsack).

2.2.2 Travelling Salesman Problem (TSP)

In the TSP, a salesman must visit each city in his designated area and then return home. In our case, the worker (tool) must perform each job and then return to the starting condition. The problem can be visualised on a graph. Each city (job) becomes a node. Arc lengths correspond to the distance between the attached cities (job changeover times). The salesman wants to find the shortest tour of the graph. A tour is a complete cycle. Starting at a home city, each city must be visited exactly one time before returning home. Each leg of the tour travels on an arc between two cities. The length of the tour is the sum of the lengths of the arcs selected. Fig 2.1 illustrates a five-city TSP. Trip lengths are shown on the arcs in Fig 2.1, the distance from city i to j is denoted by c_{ij}. We have assumed in the figure that all paths (arcs) are bi-directional. If arc lengths differ depending on the direction of the arc the TSP-formulation is said to be asymmetric, otherwise it is symmetric. A possible tour is shown in Fig 2.2. The cost of this tour is c12 + c24 + c43 + c35 + c51.

Several mathematical formulations exist for the TSP. One approach is to let x_{ij} be 1 if city j is visited immediately after i, and be 0 if otherwise. A formal statement of TSP is given as follows:

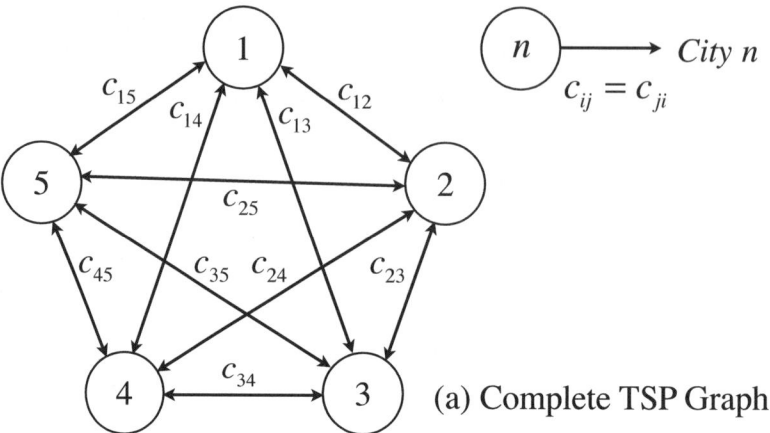

Fig. 2.1. (a) TSP illustrated on a graph

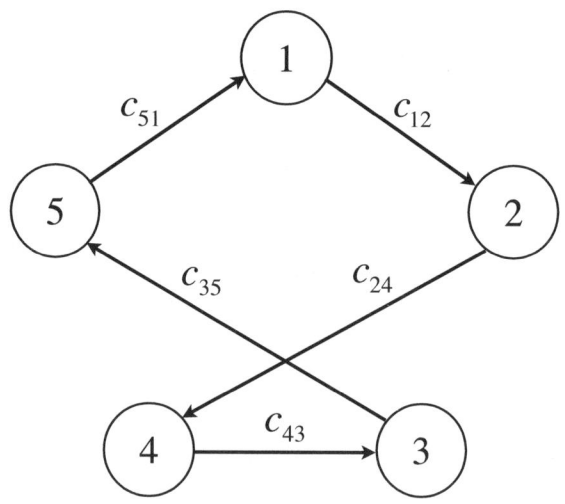

(b) One possible tour

Fig. 2.2. (b) TSP illustrated on a graph

minimise
$$\sum_{i=1}^{N}\sum_{j=1}^{N} c_{ij}x_{ij} \qquad (2.3)$$

subject to
$$\sum_{j=1}^{N} x_{ij} = 1; \forall i \qquad (2.4)$$

$$\sum_{i=1}^{N} x_{ij} = 1; \forall i \qquad (2.5)$$

No subtours

$$x_{ij} = 0 \text{ or } 1 \qquad (2.6)$$

No *subtours* mean that there is no need to return to a city prior to visiting all other cities. The objective function accumulates time as we go from city i to j. Constraint 2.4 ensures that we leave each city. Constraint 2.5 ensures that we visit (enter) each city. A subtour occurs when we return to a city prior to visiting all other cities. Restriction 6.6 enables the TSP-formulation, differ from a linear assignment programming (LAP) formulation. Unfortunately, the non-subtour constraint significantly complicates model solution. One reasonable construction procedure for solving TSP is the closest insertion algorithm. This is now discussed.

The Traveling Salesman Problem (TSP) is a fairly universal, strict-sense combinatorial problem into which many other *strict-sense* combinatorial problems can be transformed. Consequently, many findings about DE's performance on the TSP can be extrapolated to other **strict-sense** combinatorial problems.

2.2.2.1 TSP Using Closest Insertion Algorithm

The closest insertion algorithm starts by selecting any city. We then proceed through $N-1$ stages, adding a new city to the sequence at each stage. Thus a partial sequence is always maintained, and the sequence grows by one city each stage. At each stage we select the city from those currently unassigned that is closest to any city in the partial sequence. We add the city to the location that causes the smallest increase in the tour length. The closest insertion algorithm can be shown to produce a solution with a cost no worse than twice the optimum when the cost matrix is symmetric and satisfies the triangle inequality. In fact, the closest insertion algorithm may be a useful seed-solution for combinatorial search methods when large problems are solved. Symmetric implies $c_{ij} = c_{ji}$ where c_{ij} is the cost to go from city i directly to city j. Unfortunately, symmetry need not exist in our changeover problem. Normally, the triangular inequality ($c_{ij} \leq c_{ik} + c_{kj}$) will be satisfied, but this alone does not suffice to ensure the construction of a good solution. We may also try repeated application of the algorithm choosing a different starting city each time and then choose the best sequence found. Of course, this increases our workload by a factor of N. Alternatively, a different starting city may be chosen randomly for a specific number of times, less than the total number of cities. This option is preferred for large problem instances.

We now state the algorithm formally. Let S_a be the set of available (unassigned) cities at any stage. S_p will be the partial sequence in existence at any stage and is denoted $S_p = \{s_1, s_2, ..., s_n\}$, implying that city s_2 immediately follows s_1. For each unassigned city j, we use $r(j)$ to keep track of the city in the partial sequence that is closest to j. We store $r(j)$ only to avoid repeating calculations at each stage. Last, bracketed subscripts $[i]$ refer to the ith city in the current partial sequence. The steps involved are:

STEP 0. Initialize, $N = 1, S_p = \{1\}, S_a = \{2,..,N\}$. For $j = 2,...,N. r(j) = 1$
STEP 1. Select new city. Find $j^* = \arg\min_{j \in S_a} \{c_{j,c(j)}, \text{ or } c_{c(j),j}\}$. Set $n = n+1$
STEP 2. Insert j^*, update $r(j)$ $S_a = S_a - j^*$. Find City $i^* \in S_p$ such that $i^* = \arg\min_{[i] \in S_p} \{c_{[i]j^*} + c_{j^*,[i+1]} - c_{[i],[i+1]}\}$. Update $S_p = \{s_1,...,i^*,j^*,i^*+1,..,s_n\}$. For all $j \in S_a$ if $\min\{c_{j,j^*}, c_{j^*,j}\} < c_{j,r(1)}$ then $r(j) = j^*$. If $n < N$, go to 2.

As can be seen, the closest insertion algorithms a constructive method. In order to understand the steps involved, let us consider an example related to changeover times for a flexible manufacturing cell (FMC).

Example 2.1

Table 2.1 shows the changeover times for a flexible manufacturing-cell. A machine is finishing producing batch T1 and other batches are yet to be completed. We are to use the closest insertion heuristic to find a job sequence, treating the problem as a TSP.

Table 2.1. Changeover times (hrs)

From/To	T1	T2	T3	T4	T5
T1	-	8	14	10	12
T2	4	-	15	11	13
T3	12	17	-	1	3
T4	12	17	5	-	3
T5	13	18	15	2	-

Solution

Step 0 $S_p = \{1\}, S_a = \{2,3,4,5\}, r(j) = 1; j = 2,..,5$ This is equivalent to choosing
$\quad\quad\quad j \in S_a$
the first city from the partial-list and eliminating this city form the available list.

Step 1 Select the new city: find $j^* = \arg\min_{j \in S_a} \{c_{j,r[j]}\}$ and set $n = n+1$

$\min_{j \in S_a}\{c_{12}, c_{13}, c_{14}, c_{15}, c_{21}, c_{31}, c_{41}, c_{51}\} = \min\{8, 14, 10, 12, 0, 0, 0, 0\} = 8 \,; j^* = 2$. But ignore c_{21}, c_{31}, c_{41} and c_{51} because city 1 is already considered in S_a.

Step 2 Insert city 2, and update $r(j)$ for the remaining jobs 3, 4, and 5 $S_p = \{1,2\}; S_a = \{3,4,5\}, c_{12} + c_{21} - c_{14} = 8 + 4 - 0 = 12$

Step 1 Select new city:
$\min\{c_{23}, c_{24}, c_{25}, c_{32}, c_{42}, c_{52}\} = \{15, 11, 13, 17, 17, 18\} = 11; j^* = 4; n = 3$. So we have job 4 after job 1 or 2.

Step 2 Insert job 4 There are the following possibilities from $\{1,2\}$: $\{1,2,4\}$ or $\{1,4,2\}$
For $\{1,2,4\}, c_{12} + c_{24} - c_{14} = 8 + 11 - 10 = 9$
For $\{1,4,2\}, c_{14} + c_{42} - c_{12} = 10 + 17 - 8 = 19$
The minimum occurs for inserting job 2 after job 4. Update $r(j)$ for remaining jobs 3, 5 i.e., r(3) = r(5) = 4

Step 1 Select new job.
$$\min_{j \in S_a} \{c_{34}, c_{54}, c_{43}, c_{45}\} = \min\{1, 2, 5, 3\}; j^* = 3$$
$\min = c_{34}$ but 4 is already considered. Hence, $j^* = 3$.

Step 2 Insert job 3
There are the following possibilities from $\{1,2,4\}$:
$\{1,2,4,3\} : c_{43} + c_{31} - c_{41} = 5 + 12 - 12 = 5$
$\{1,2,3,4\} : c_{23} + c_{34} - c_{24} = 15 + 1 - 11 = 5$
$\{1,3,2,4\} : c_{13} + c_{32} - c_{12} = 14 + 17 - 8 = 25$
Choosing $\{1,2,4,3\}$ breaks the tie. Updating $r[5] = 3$

Step 1 Select new job.
Since job 5 remains, $j^* = 5$

Step 2 Insert job 5
There are the following possibilities from $\{1,2,4,3\}$
$\{1,2,4,3,5\} : c_{35} + c_{51} - c_{31} = 3 + 13 - 12 = 4$
$\{1,2,3,4,3\} : c_{45} + c_{53} - c_{43} = 3 + 15 - 5 = 13$
$\{1,3,2,4,3\} : c_{25} + c_{54} - c_{24} = 13 + 2 - 11 = 4$
$\{1,5,2,4,3\} : c_{15} + c_{52} - c_{12} = 12 + 18 - 8 = 22$
Choosing $\{1,2,4,3,5\}$ breaks the tie as the final sequence. The cost $= c_{12} + c_{24} + c_{43} + c_{35} = 8 + 11 + 5 + 3 = 27$

The TSP construction is shown in Fig 2.3. The meaning of this solution is that batch *1* is first produced, followed by batch *2*, then batch *4*, then batch *3*, and finally batch *5*.

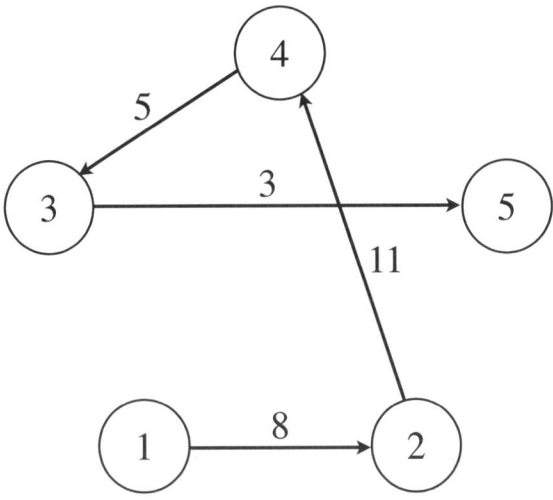

Fig. 2.3. TSP solution for Example 2.1

2.2.3 Automated Drilling Location and Hit Sequencing

Consider an automated drilling machine that drills holes on a flat sheet. The turret has to be loaded with all the tools required to hit the holes. There is no removing or adding of tools. The machine bed carries the flat plate and moves from home, locating each scheduled hit on the flat plate under the machine turret. Then the turret rotates and aligns the proper tool under the hit. This process continues until all hits are completed and the bed returns to the home position.

There are two problems to be solved here. One is to load tools to the turrets and the other is to locate or sequence hits. The objective is to minimize the cycle time such that the appropriate tools are loaded and the best hits-sequence is obtained. The problem can therefore be divided into two: (i) solve a TSP for the inter−hit sequencing; (ii) solve a quadratic assignment problem (QAP) for the tool loading. [21] developed a mathematical formulation to this problem and iterated between the TSP and QAP. Once the hit sequence is known, the sequence of tools to be used is then fixed since each hit requires specific tool. On the other hand, if we know the tool assignment on the turret, we need to know the inter-hit sequence. Connecting each hit in the best sequence is definitely a TSP, where we consider the machine bed home as the home for the TSP, and each hit, a city. Inter−hit travel times and the rotation of the turret are the costs involved and we take the maximum between them, i.e. inter−hit cost = max (inter−hit travel time, turret rotation travel time). The cost to place tool k in position i and tool l in position j is the time it takes the turret to rotate from i to j multiplied by the number of times the turret switches from tool k to l.

The inter-hit travel times are easy to estimate from the geometry of the plate to be punched and the tools required per punch. The inter−hits times are first estimated and then adjusted according to the turret rotation times. This information constitutes the data for solving the TSP. Once the hit sequence is obtained from the TSP, the tools are placed, by solving the QAP. Let us illustrate the TSP−QAP solution procedure by considering an example.

Example 2.2

A numerically controlled (NC) machine is to punch holes on a flat metal sheet and the hits are shown in Fig 2.4. The inter-hit times are shown in Table 2.2. There are four tools $\{a,b,c,d\}$ and the hits are $\{1,2,3,4,5,6,7\}$. The machine turret can hold five tools and rotates in clockwise or anti-clockwise direction. When the turret rotates from one tool position to an adjacent position, it takes *60* time units. It takes *75* time units and *90* time units to two locations and three locations respectively. The machine bed home is marked *0*. Assign tools to the turret and sequence the hits.

Solution

From the given inter−hit times, modified inter-hit times have to be calculated using the condition: inter−hit cost = max (inter-hit travel time, turret rotation travel time). For example, for inter−hit between locations *1* and *2*, the inter-hit travel time is *50* time units. Now, the tool for hit *1* is c while the tool for hit 2 is a. This means there is

Table 2.2. Inter–hit travel times

Hit	0	1	2	3	4	5	6	7
0	-	50	100	50	100	150	100	200
1	50	-	50	100	50	100	150	150
2	100	50	-	150	100	50	200	100
3	50	100	150	-	50	100	50	150
4	100	50	100	50	-	50	100	100
5	150	100	50	100	50	-	150	50
6	100	150	200	50	100	150	-	100
7	200	150	100	150	100	50	100	-

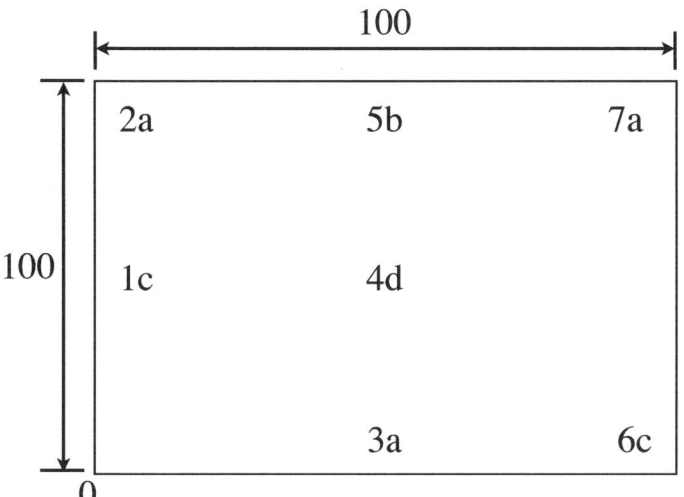

Fig. 2.4. Flat metal sheet to be punched

change in tools because that the turret will rotate. The cost of rotation is *60* time units, which exceeds the *50* inter-hit time unit. This means that the modified inter–hit time between locations *1* and *2* is *60* time units. From the home to any hit is not affected. The modified inter–hit times are shown in Table 2.3. This information is used for TSP. One TSP solution for Table 2.2 is $\{0,1,2,3,4,5,6,7\}$, with a cost of *830*. We used the DE heuristic to obtain tool sequence of $c \to d \to b \to a \to c \to a$, and the cost is *410*. As can be seen a better solution is obtain by the latter. Let us explain how we obtained the tool sequence. Solving the TSP using DE, the sequence obtaineed is $\{2,5,6,8,7,4,1,3\}$ or $\{1,4,5,7,6,3,0,2\}$. What we do is to refer to Fig 2.4 and get the labels that corrspond to this sequence as $\{c,d,b,a,c,a,a\}$. Hence the optimum sequence is $c-d-b-a-c$.

Table 2.3. Modified inter–hit travel times (considering turret movements)

Hit	Hit							
	0	1	2	3	4	5	6	7
0	-	50	100	50	100	150	100	200
1	50	-	60	100	60	100	150	150
2	100	60	-	150	100	60	200	100
3	50	100	150	-	60	100	60	150
4	100	60	100	60	-	60	100	100
5	150	100	60	100	60	-	150	60
6	100	150	200	60	100	150	-	100
7	200	150	100	150	100	60	100	-

2.2.4 Dynamic Pick and Place (DPP) Model of Placement Sequence and Magazine Assignment

Products assembled by robots are typical in present manufacturing system. To satisfy growing large scale demand of products efficient methods of product assemble is essential to reduce time frame and maximize profit. The Dynamic Pick and Place (DPP) model of Placement Sequence and magazine Assignment (SMA) is an interesting problem that could be solved using standard optimizing techniques, such as discrete or permutative DE. DPP model is a system consists of robot, assemble board and magazine feeder which move together with different speeds and directions depends on relative distances between assemble points and also on relative distances between magazine components. Major difficulty to solve this problem is that the feeder assignment depends on assembly sequence and vice versa. Placement sequence and magazine assignment (SMA) system has three major components robot, assembly board and component slots. Robot picks components from horizontal moving magazine and places into the predefined positions in the horizontal moving assembly board. To optimize production time frame assembly sequence and feeder assignment need to be determined. There are two models for this problem: Fixed Pick and place model and Dynamic Pick and Place model. In the FPP model, the magazine moves in x direction only while the board moves in both x-y directions and the robot arm moves between fixed "pick" and "place" points. In the DPP model, both magazine and board moves along x-axis while the robot arm moves between dynamic "pick" and "place" points. See Fig 2.5. Principal objective is to minimize total tardiness of robot movement hence minimize total assembly time.

There are few researchers who had solved the assembly sequence and feeder assignment problem by the DPP model. This is because this problem is quite challenging. [12] had proved that DPP has eliminated the robot waiting time by the FPP model. [11] used simulated annealing algorithm and obtained solutions better than previous approaches but the computation efficiency was quite low. Wang et al. have developed their own heuristic approach to come up with some good solutions. [19] proposed a new heuristic to improve Wang's approach based on the fact that assembly time depends on the relative position of picking points as well as placement points. The main objective of DPP model is to eliminate the robot waiting time. To avoid tardiness robot arm tends to

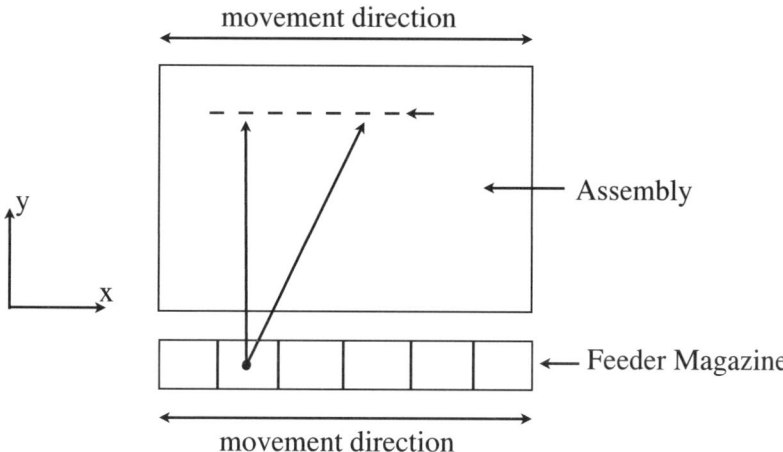

Fig. 2.5. Robot movement

move in shortest possible path (i.e. always tries to move in vertical direction) if vertical movement is not possible than it needs to stretch/compress its arm to avoid tardiness. This section reveals the formulation of DPP problem statement using the following notations:

V_a	speed of assembly arm
V_b	speed of board
$V\neg_m$	speed of magazine
N	number of placement components
K	number of component types ($K \leq N$)
$b(i)$	i^{th} placement in a placement sequence
$m(i)$	i^{th} placement in a pick sequence
$x_{i+1}^m = M_{i+1}^1 + M_{i+1}^2$	interception distance of robot arm and magzine
$x_{i+1}^b = B_{i+1}^1 + B_{i+1}^2$	interception distance of robot arm and board
$T(m(i), b(i))$	robot arm travel time from magazine location $m(i)$ to board location $m(i)$.
$T(b(i), b(i))$	robot arm travel time from board location $b(i)$ to magazine location $m(i)$.
$T place$	time taken to place the component
$T pick$	time taken to pick the component
CT	total assembly time

Fig 2.6 shows possible movements of board and magazine in DPP model [19]. Suppose the robot arm has finished placing the ith component at point $B(i)$ then moves to pick the next $(i+1)^{th}$ component from slot $M(i+1)$ on the magazine. If magazine is able to travel distance $d(a,c) = ||a - c||$ before the robot actually arrives vertically towards

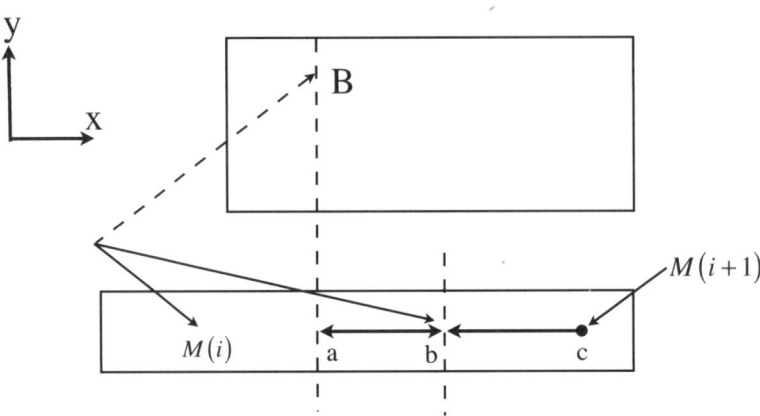

Fig. 2.6. DPP model

magazine then no interception will occur, but if the magazine fails then robot has to compress/stretch (intercept) its arm in x-direction of distance $d(a,b) = ||a-b||$ to get the component without waiting for the component $M(i+1)$ to arrive at point a.

Fig 2.6 shows that slot $M(i+1)$ at point c has to reach at point a through travelling distance $d(a,c) = ||a-c||$ to avoid robot interception. Suppose magazine only managed to travel distance $d(b,c) = ||b-c||$ before the robot reached to the magazine i.e. $T(M(i), b(i)) + T_{place} + \frac{y_i}{v_r} \geq \frac{d(a,c)}{v_m}$. Then the robot has to stretch its arm to get to the point b. Hence tardiness is eliminated by robot interception. Exactly same movement principles are applied when robot arm moves from magazine to board.

Where total assembly time CT need to be minimized subject to constraints:

1. Board to magazine

$$f(T(m(i),b(i)) + T_{place} + \frac{y_i}{v_r} \geq \frac{d(a,c)}{v_m}) \; Then$$
$$T(b(i), m(i+1)) = \frac{y_i}{v_m}$$
$$else$$
$$T(b(i), m(i+1)) = \frac{\sqrt{(y_i^2 + (x_{i+1}^m)^2)}}{v_r}$$
$$Endif$$

2. Magazine to board

$$if\,(T(b(i), m(i+1)) + T_{pick} + \frac{y_i}{v_r} \geq \frac{d(a,c)}{vb}) \; Then$$
$$T(m(i+1), b(i+1)) = \frac{y_i}{v_m}$$
$$else$$
$$T(m(i+1), b(i+1)) = \frac{\sqrt{(y_{i+1}^2 + (x_{i+1}^b)^2)}}{v_r}$$
$$Endif$$

The formulation of this problem shows that DPP model is a function of ith placement in a sequence $b(i)$, and i^{th} component in a pick sequence, $m(i)$. This is obviously a permutative-based combinatorial optimization problem which is challenging to solve.

2.2.5 Vehicle Routing Problem

Vehicle routing problem is delivery of goods to customers by a vehicle from a depot (see Fig 2.7). The goal here is to minimize the travelling distance and hence save cost. Here too an objective function would be created and inserted into the optimizer in order to obtain the best travelling path for which the cost is minimized. The application of vehicle routing problem can be applied in many places. One example is bin–picking problem. In some countries, the City Council bears a lot of extra costs on bin-picking vehicle by not following shortest path.

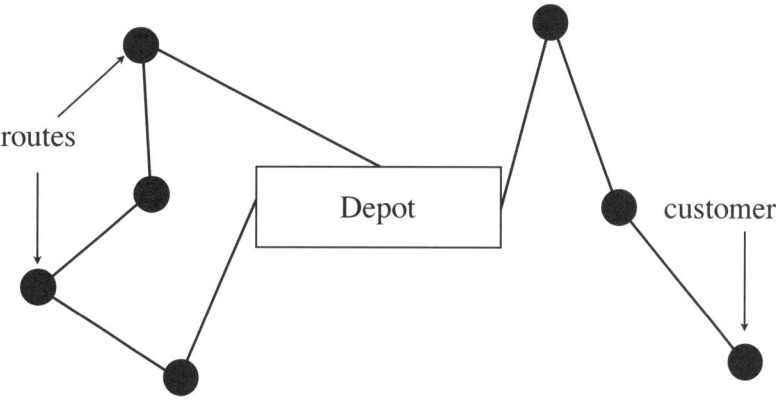

Fig. 2.7. DPP model

The CVRP is described as follows: n customers must be served from a unique depot. Each customer asks for a quantity for quantity q_i (where $i = 1,2,3,...,n$) of goods and a vehicle of capacity Q is available for delivery. Since the vehicle capacity is limited, the vehicle has to periodically return to the depot for reloading. Total tour demand is at most Q (which is vehicle capacity) and a customer should be visited only once [5].

2.2.6 Facility Location Problem

In facility location problem, we are given n potential facility location and m customers that must be served from these locations. There is a fixed cost c_j of opening facility j. There is a cost d_{ij} associated with serving customer i from facility j. We then have two sets of binary variables which are y_j is *1* if facility j is opened, *0* otherwise x_{ij} is *1* if customer i is served by facility j, *0* otherwise.

Mathematically the facility location problem can be formulated as

$$\min \sum_{j=1}^{n} c_j y_j + \sum_{i=1}^{m} \sum_{j=1}^{n} d_{ij} x_{ij}$$
$$s.t. \sum_{j=1}^{n} x_{ij} = K \quad \forall i \qquad (2.7)$$
$$x_{ij} \le y_j \quad \forall i,j$$
$$x_{ij}, y_j \in \{0, i\} \, \forall i,j$$

2.3 Permutation-Based Combinatorial Approaches

This section describes two permutation–based combinatorial DE approaches which were merely described in [10] and three other permutation–based combinatorial DE approaches which are detailed in this book.

2.3.1 The Permutation Matrix Approach

The permutation matrix approach is the idea of Price, but Storn did the experiments that document its performance [10]. The permutative matrix approach is based on the idea of finding a permutative matrix that relates two vectors. For example, given two vectors x_{r1} and x_{r1} defined in Equation 2.8:

$$x_{r1} = \begin{pmatrix} 1 \\ 3 \\ 4 \\ 5 \\ 2 \end{pmatrix}, \quad x_{r2} = \begin{pmatrix} 1 \\ 4 \\ 3 \\ 5 \\ 2 \end{pmatrix}; \qquad (2.8)$$

These two vectors encode tours, each of which is a permutation. The *permutation matrix*, **P**, that x_{r1} and x_{r1} is defined as:

$$x_{r2} = P.x_{r1}, \text{ with } P = \begin{pmatrix} 1 & 0 & 0 & 0 & 0 \\ 0 & 0 & 1 & 0 & 0 \\ 0 & 1 & 0 & 0 & 0 \\ 0 & 0 & 0 & 1 & 0 \\ 0 & 0 & 0 & 0 & 1 \end{pmatrix} \qquad (2.9)$$

```
for(i = 1; i < M; i++) //search all columns of P
{
  if(elementp(i,i) of P is 0) // 1 not on diagonal
  {
    if(rand() > δdel) //if random number ex [0,1] exceeds δdel
    {
      j = 1; // find row where p(j,i) = 1
      while(p(j,i)! = 1)j++;
    }
  }
}
```

Fig. 2.8. Algorithm to apply the factor δ to the difference permutation, **P**

Price gives an algorithm that scales the effect of the permutation matrix as shown in Fig 2.8.

2.3.2 Adjacency Matrix Approach

Storn developed the adjacency matrix approach outlined in this section [10]. There are some rules that govern the adjacency matrix approach. When tours are encoded as city vectors, the difference between rotated but otherwise identical tours is never zero. Rotation, however, has no effect on a tour's representation if it is encoded as an adjacency matrix. Storn defined the notation

$$(x+y) \bmod 2 = x \oplus y \tag{2.10}$$

which is shorthand for modulo 2 addition, also known as the "exclusive or" logical operation for the operation of the matrices. The difference matrix Δ_{ij},

$$\Delta_{ij} = A_i \oplus A_j \tag{2.11}$$

is defined as the analog of DE's traditional difference vector. Consider for example, the valid TSP matrices A_1 and A_2,

$$A_1 = \begin{pmatrix} 0 & 1 & 0 & 0 & 1 \\ 1 & 0 & 0 & 1 & 0 \\ 0 & 0 & 0 & 1 & 1 \\ 0 & 1 & 1 & 0 & 0 \\ 1 & 0 & 1 & 0 & 0 \end{pmatrix}, \quad A_2 = \begin{pmatrix} 0 & 1 & 0 & 0 & 1 \\ 1 & 0 & 1 & 0 & 0 \\ 0 & 1 & 0 & 1 & 0 \\ 0 & 0 & 1 & 0 & 1 \\ 1 & 0 & 0 & 1 & 0 \end{pmatrix}, \tag{2.12}$$

and their difference given as

$$\Delta_{1,2} = A_1 \oplus A_2 = \begin{pmatrix} 0 & 0 & 0 & 0 & 0 \\ 0 & 0 & 1 & 1 & 0 \\ 0 & 1 & 0 & 0 & 1 \\ 0 & 1 & 0 & 0 & 1 \\ 0 & 0 & 1 & 1 & 0 \end{pmatrix} \tag{2.13}$$

From the definition of A_1 there are 1's in column 1 in rows {2 and 5}, in column 2 there are 1's in rows {1 and 4}, in column 3 there are 1's in rows {4 and 5}, in column 4 there are 1's in rows {2 and 3}, in column 5 there are 1's in rows {1 and 3} respectively. These pair–wise numbers define the adjacency relationships. Considering {2 and 5} and {4 and 5} it is shown that '5' is common and {2 and 4} are adjacent. Considering {1 and 4} and {1 and 3} it is shown that '1' is common and {1 and 3} are adjacent. Continuing in this manner it could be observed that Fig 2.9 shows the graphical interpretation of A_1, A_2 and Δ_{ij}.

2.3.3 Relative Position Indexing

In the relative position indexing approach [4], permutations are obtained by determining the relative sizes of the different parameters defining an instance. Let there be an

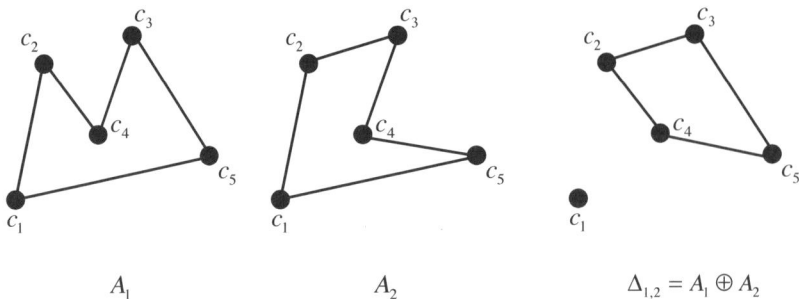

Fig. 2.9. Graphical interpretations of A_1, A_2 and the difference matrix Δ_{ij}

instance of four cities that define a tour such that these are initially generated by DE as $x_{1,f} = \{0.5\ 0.8\ 0.2\ 0.6\}$. Another instance of such four cities could simple be simply defined as $x_{2,f} = \{0.6\ 0.1\ 0.3\ 0.4\}$. In relative indexing, these instances encode permutations given as $x_1 = \{2\ 4\ 1\ 3\}$ and $x_2 = \{4\ 1\ 2\ 3\}$ respectively. In the first case for example the lowest value which is 0.2 is in the third position so it is allocated a label of 1; the next higher value is 0.5 which occupies the first position and it is allocated the label 2 and so on. Let there be a third instance denoted as $x_{3,f} = \{0.6\ 0.8\ 0.3\ 0.5\}$. Then we have $x_3 = \{3\ 4\ 1\ 2\}$. The concept is fairly simple. The subscript f indicates floating point.

The basic idea behind DE is that two vectors define a difference that can then be added to another vector as a mutation. The same idea transfers directly to the realm of permutations, or the permutation group. Just as two vectors in real space define a difference vector that is also a vector, two permutations define a mapping that is also a permutation. Therefore, when mutation is applied to with F = 0.5, the floating-point mutant vector, v_f, is

$$v_f = x_{r3,f} + F\left(x_{r1,f} - x_{r2,f}\right)$$
$$= \{0.6\ 0.8\ 0.3\ 0.5\} + 0.5\{-0.1\ 0.7\ -0.1\ 0.2\} \quad (2.14)$$
$$= \{0.55\ 1.15\ 0.25\ 0.6\}$$

The floating-point mutant vector, v_f, is then transformed back into the integer domain by assigning the smallest floating value (0.25) to the smallest integer (1), the next highest floating value (0.55) to the next highest integer (2), and so on to obtain $v = \{2\ 4\ 1\ 3\}$. [10] noted that this backward transformation, or "relative position indexing", always yields a valid tour except in the unlikely event that two or more floating–point values are the same. When such an event occurs, the trial vector must be either discarded or repaired.

2.3.4 Forward/Backward Transformation Approach

The forward/backward transformation approach is the idea of [6], and is generally referred to as Onwubolu's approach [10]. There are two steps involved:

Forward Transformation

The transformation scheme represents the most integral part of the code. [6] developed an effective routine for the conversion of permutative-based indices into the continuous domain. Let a set of integer numbers be represented as $\mathbf{x}_i \in \mathbf{x}_{i,G}$ which belong to solution $x_{j,i,G=0}$. The formulation of the forward transformation is given as:

$$\mathbf{x}'_i = -1 + \alpha \mathbf{x}_i \tag{2.15}$$

where the value α is a small number.

Backward Transformation

The reverse operation to forward transformation, converts the real value back into integer as given in 2.16 assuming $\mathbf{x}'$ to be the real value obtained from 2.15.

$$\text{int}[\mathbf{x}_i] = (1 + \mathbf{x}'_i)/\alpha \tag{2.16}$$

The value $\mathbf{x}_i$ is rounded to the nearest integer. [9], [2, 3] have applied this method to an enhanced DE for floor shop problems.

2.3.5 Smallest Position Value Approach

The smallest position value (SPV) approach is the idea of [20] in which a unique solution representation of a continuous DE problem formulation is presented and the SPV rule is used to determine the permutations. Applying this concept to the GTSP, in which a tour is required, integer parts of the parameter values (s_j) in a continuous DE problem formulation represent the nodes (v_j). Then the random key values (s_j) are determined by simply subtracting the integer part of the parameter x_j from its current value considering the negative signs, i.e., $s_j = x_j - \text{int}(x_j)$. Finally, with respect to the random key values (s_j), the smallest position value (SPV) rule of [20] is applied to the random key vector to determine the tour π. They adapted the encoding concept of [1] for solving the GTSP using GA approach, where each set V_j has a gene consisting of an integer part between $[1, |V_j|]$ and a fractional part between $[0, 1]$. The integer part indicates which node from the cluster is included in the tour, and the nodes are sorted by their fractional part to indicate the order. The objective function value implied by a solution x with m nodes is the total tour length, which is given by

$$F(\pi) = \sum_{j=1}^{m-1} d_{\pi_j \pi_{j+1}} + d_{\pi_m \pi_1} \tag{2.17}$$

$V = \{1,..,20\}$ and $V_1 = \{1,..,5\}$, $V_2 = \{6,,..,10\}$, $V_3 = \{11,..,15\}$ and $V_4 = \{16,..,20\}$. Table 2.4 shows the solution representation of the DE for the GTSP.

In Table 2.4, noting that $[1, |V_j|]$, the integer parts of the parameter values (s_j) are respectively decoded as $\{4, 3, 1, 3\}$. These decoded values are used to extract the nodes from the clusters V_1, V_2, V_3, V_4. The first node occupies the fourth position in V_1, the second node occupies the third position in V_2, the third node occupies the first position in

Table 2.4. SPV Solution Representation

j	1	2	3	4
x_j	4.23	-3.07	1.80	3.76
v_j	4	8	11	18
s_j	0.23	-0.07	0.80	0.76
π_j	8	4	18	11
$F(\pi)$	$d_{8,4}$	$d_{4,18}$	$d_{18,11}$	$d_{11,8}$

V_3, while the fourth node occupies the third position in V_4. Extracting these labels show that the nodes are $\{4,8,11,18\}$ The random key values are $\{0.23,-0.07,0.80,0.76\}$; finally, with respect to the random key values (s_j), the smallest position value (SPV) rule is applied to the random key vector by arranging the values in a non-descending order $\{-0.07,0.23,0.76,0.08\}$ to determine the tour $\pi\{8,4,18,11\}$. Using equation 2.17, the total tour length is then obtained as

$$F(\pi) = \sum_{j=1}^{m-1} d_{\pi_j \pi_{j+1}} + d_{\pi_m \pi_1} = d_{8,4} + d_{4,18} + d_{18,11} + d_{11,8}$$

In this approach, a problem may rise such that when the DE update equations are applied, any parameter value might be outside of the initial search range, which is restricted to the size of each cluster. Let $x_{\min}[j]$ and $x_{\max}[j]$ represent the minimum and maximum value of each parameter value for dimension j. Then they stand for the minimum and maximum cluster sizes of each dimension j. Regarding the initial population, each parameter value for the set V_j is drawn uniformly from $[-V_j+1, V_j+1]$. Obviously, $x_{\max}[j]$ is restricted to $[V_j+1]$, whereas $x_{\min}[j]$ is restricted to $-x_{\max}[j]$. During the reproduction of the DE, when any parameter value is outside of the cluster size, it is randomly reassigned to the corresponding cluster size again.

2.3.6 Discrete/Binary Approach

Tasgetiren et al. present for the first time in this chapter, the application of the DDE algorithm to the GTSP. They construct a unique solution representation including both cluster and tour information is presented, which handles the GTSP properly when carrying out the DDE operations. The Population individuals can be constructed in such a way that first a permutation of clusters is determined randomly, and then since each cluster contains one or more nodes, a tour is established by randomly choosing a single node from each corresponding cluster. For example, n_j stands for the cluster in the jth dimension, whereas π_j represents the node to be visited from the cluster n_j.

Now, consider a GTSP instance with $N = \{1,..,25\}$ where the clusters are $n_1 = \{1,...,5\}$, $n_2 = \{6,...,10\}$, $n_3 = \{11,...,15\}$, $n_4 = \{16,...,20\}$ and $n_5 = \{21,...,25\}$. Table 2.5 shows the discrete/binary solution representation of the DDE for the GTSP.

A permutation of clusters is determined randomly as $\{4,1,5,2,3\}$. This means that the first node is randomly chosen from the fourth cluster (here 16 is randomly chosen); the second node is randomly chosen from the first cluster (here 5 is randomly chosen);

Table 2.5. Discrete/binary Solution Representation

	j	1	2	3	4	5
	n_j	4	1	5	2	3
X	π_j	16	5	22	8	14
	$d_{\pi_j \pi_{j+1}}$	$d_{16,5}$	$d_{5,22}$	$d_{22,8}$	$d_{8,14}$	$d_{14,16}$

the third node is randomly chosen from the fifth cluster (here 22 is randomly chosen); the fourth node is randomly chosen from the second cluster (here 8 is randomly chosen); and the fifth node is randomly chosen from the third cluster (here 14 is randomly chosen).

As already illustrated, the objective function value implied by a solution x with m nodes is the total tour length, which is given by:

$$F(\pi) = \sum_{j=1}^{m-1} d_{\pi_j \pi_{j+1}} + d_{\pi_m \pi_1} \qquad (2.18)$$

This leads to the total tour length being obtained as

$$F(\pi) = \sum_{j=1}^{m-1} d_{\pi_j \pi_{j+1}} + d_{\pi_m \pi_1} = d_{16,5} + d_{5,22} + d_{22,8} + d_{8,14} + d_{14,16}$$

2.3.7 Discrete Set Handling Approach

Discrete set handling is an algorithmic approach how to handle in a numerical way objects from discrete set. Discrete set usually consist of various elements with non-numerical nature. In its canonical form DE is only capable of handling continuous variables. However extending it for optimization of discrete variables is rather easy. Only a couple of simple modifications are required. In evolution instead of the discrete value x_i itself, we may assign its index, i, to x. Now the discrete variable can be handled as an integer variable that is boundary constrained to range $< 1, 2, 3, ..., N >$. So as to evaluate the objective function, the discrete value, x_i, is used instead of its index i. In other words, instead of optimizing the value of the discrete variable directly, we optimize the value of its index i. Only during evaluation is the indicated discrete value used. Once the discrete problem has been converted into an integer one, the methods for handling integer variables can be applied. The principle of discrete parameter handling is depicted in chapter 7.3.

2.3.8 Anatomy of Some Approaches

[10] carried out anatomy of the four permutation–based combinatorial DE approaches described in their book (see Table 2.6). This exercise excludes smallest position value, discrete/binary and discrete set handling approaches.

Table 2.6. Anatomy of four permutation-based combinatorial DE approaches[+]

Approach	Observations
Permutation Matrix	In practice, this approach tends to stagnate because moves derived from the permutation matrix are seldom productive. In addition, this method is unable to distinguish rotated but otherwise equal tours. Because they display a unique binary signature, equal tours can be detected by other means, although this possibility is not exploited in the algorithm described in Fig 2.8.
Adjacency Matrix	(1) This scheme preserves good sections of the tour if the population has almost converged, i.e., if most of the TSP matrices in the population contain the same sub−tours. When the population is almost converged, there is a high probability that the difference matrix will contain just a few ones, which means that there are only a few cities available for a 2−exchange.
Relative Position Indexing	(1) This approach resembles traditional DE because they both use vector addition, although their ultimate effect is to shuffle values between parameters, i.e., generate permutations. (2) This approach impedes DE's self-steering mechanism because it fails to recognize rotated tours as equal. (3) A closer look, however, reveals that DE's mutation scheme together with the forward and backward transformations is, in essence, a shuffling generator. (4) In addition, this approach does not reliably detect identical tours because the difference in city indices has no real significance. For example, vectors with rotated entries, e.g., $(2,3,4,5,1)$ and $(1,2,3,4,5)$, are the same tour, but their difference, e.g., $(1,1,1,1,-4)$, is not zero.
Forward/backward Transformation	(1) This approach resembles traditional DE because they both use vector addition, although their ultimate effect is to shuffle values between parameters, i.e., generate permutations. (2) This approach impedes DE's self−steering mechanism because it fails to recognize rotated tours as equal. (3) In addition, Onwubolu's method usually generates invalid tours that must be repaired. Even though competitive results are reported in Onwubolu there is reason to believe that the success of this approach is primarily a consequence of prudently chosen local heuristics and repair mechanisms, not DE mutation.

[+] Described in Price et al. (2005).

2.4 Conclusions

There has been some reservation that although DE has performed well on wide−sense combinatorial problems, its suitability as a combinatorial optimizer is still a topic of considerable debate and a definitive judgment cannot be given at this time. Moreover, it is said that although the DE mutation concept extends to other groups, like the permutation group, there is no empirical evidence that such operators are particularly effective. The opinion expressed in this book is similar to that of [18] that all good heuristics are able to transform a combinatorial problem into a space which is amenable for search, and that there is no such thing as an "all-cure" algorithm for combinatorial problems. For example, particle swarm optimization (PSO) works fairly well for combinatorial problems, but only in combination with a good tailored heuristic. If such a heuristic is used, then PSO can locate promising regions. The same logic applies to a number of optimization approaches.

While the anatomy described in Table 2.6 favors the adjacency matrix and permutation matrix approaches, compared to the forward/backward transformation relative position indexing approaches, it is not known in the literature where the adjacency matrix and permutation matrix approaches have been applied to real–life permutation-based combinatorial problems.

In this book, it is therefore concluded that:

1. The original classical DE which Storn and Price developed was designed to solve only problems characterized by continuous parameters. This means that only a subset of real-world problems could be solved by the original canonical DE.
2. For quite some time, this deficiency made DE not to be employed to a vast number of real-world problems which characterized by permutative-based combinatorial parameters.
3. This book complements that of [10] and vice versa. Taken together therefore, both books will be needed by practitioners and students interested in DE in order to have the full potentials of DE at their disposal. In other words, DE as an area of optimization is incomplete unless it can deal with real–life problems in the areas of continuous space as well as permutative-based combinatorial domain.

References

1. Bean, J.: Genetic algorithms and random keys for sequencing and optimization. ORSA, Journal on Computing 6, 154–160 (1994)
2. Davendra, D., Onwubolu, G.: Flow Shop Scheduling using Enhanced Differential Evolution. In: Proceeding of the 21st European Conference on Modelling and Simulation, Prague, Czech Republic, June 4-5, pp. 259–264 (2007)
3. Davendra, D., Onwubolu, G.: Enhanced Differential Evolution hybrid Scatter Search for Discrete Optimisation. In: Proceeding of the IEEE Congress on Evolutionary Computation, Singapore, September 25-28, pp. 1156–1162 (2007)
4. Lichtblau, D.: Discrete optimization using Mathematica. In: Callaos, N., Ebisuzaki, T., Starr, B., Abe, M., Lichtblau, D. (eds.) World multi–conference on systemics, cybernetics and informatics (SCI 2002), International Institute of Informatics and Systemics, vol. 16, pp. 169–174 (2002) (Cited September 1, 2008),
   http://library.wolfram.comlinfocenter/Conferences/4317
5. Mastolilli, M.: Vehicle routing problems (2008) (Cited September 1, 2008),
   http://www.idsia.ch/~monaldo/vrp.net
6. Onwubolu, G.: Optimisation using Differential Evolution Algorithm. Technical Report TR-2001-05, IAS (October 2001)
7. Onwubolu, G.: Optimizing CNC drilling machine operation: traveling salesman problem-differential evolution approach. In: Onwubolu, G., Babu, B. (eds.) New optimization techniques in engineering, pp. 537–564. Springer, Heidelberg (2004)

8. Onwubolu, G., Clerc, M.: Optimal path for automated drilling operations by a new heuristic approach using particle swarm optimization. Int. J. Prod. Res. 42(3), 473–491 (2004)
9. Onwubolu, G., Davendra, D.: Scheduling flow shops using differential evolution algorithm. Eur. J. Oper. Res. 171, 674–679 (2006)
10. Price, K., Storn, R., Lampinen, J.: Differential Evolution. Springer, Heidelberg (2005)
11. Su, C., Fu, H.: A simulated annealing heuristic for robotics assembly using the dynamic pick–and–place model. Prod. Plann. Contr. 9(8), 795–802 (1998)
12. Su, C., Fu, H., Ho, L.: A novel tabu search approach to find the best placement sequence and magazine assignment in dynamic robotics assembly. Prod. Plann. Contr. 9(6), 366–376 (1998)
13. Storn, R.: Differential evolution design of an R–filter with requirements for magnitude and group delay. In: IEEE international conference on evolutionary computation (ICEC 1996), pp. 268–273. IEEE Press, New York (1996)
14. Storn, R.: On the usage of differential evolution for function optimization. In: NAFIPS, Berkeley, pp. 519–523 (1996)
15. Storn, R.: System design by constraint adaptation and differential evolution. IEEE Trans. Evol. Comput. 3(I), 22–34 (1999)
16. Storn, R.: Designing digital filters with differential evolution. In: Come, D., et al. (eds.), pp. 109–125 (1999)
17. Storn, R.: (2000) (Cited September 1, 2008), http://www.icsi.berkeley.edu/~stornlfiwiz.html
18. Storn, R., Price, K.: Differential evolution–a simple and efficient heuristic for global optimization over continuous spaces. J. Global Optim. 11, 341–359 (1997)
19. Tabucanon, M., Hop, N.: Multiple criteria approach for solving feeder assignment and assembly sequencing problem in PCB assembly. Prod. Plann. Contr. 12(8), 736–744 (2001)
20. Tasgetiren, M., Sevkli, M., Liang, Y.-C., Gencyilmaz, G.: Particle Swarm Optimization Algorithm for the Single Machine Total Weighted Tardiness Problem. In: The Proceeding of the World Congress on Evolutionary Computation, CEC 2004, pp. 1412–1419 (2004)
21. Walas, R., Askin, R.: An algorithm for NC turret punch tool location and hit sequencing. IIE Transactions 16(3), 280–287 (1984)

3
Forward Backward Transformation

Donald Davendra[1] and Godfrey Onwubolu[2]

[1] Tomas Bata University in Zlin, Faculty of Applied Informatics, Nad Stranemi 4511, Zlin 76001, Czech Republic
davendra@fai.utb.cz
[2] Knowledge Management & Mining, Inc., Richmond Hill, Ontario, Canada
onwubolu_g@dsgm.ca

Abstract. Forward Backward Transformation and its realization, Enhanced Differential Evolution algorithm is one of the permutative versions of Differential Evolution, which has been developed to solve permutative combinatorial optimization problems. Novel domain conversions routines, alongside special enhancement routines and local search heuristic have been incorporated into the canonical Differential Evolution in order to make it more robust and effective.

Three unique and challenging problems of Flow Shop Scheduling, Quadratic Assignment and Traveling Salesman have been solved, utilizing this new approach. The promising results obtained have been compared and analysed against other benchmark heuristics and published work.

3.1 Introduction

Complexity and advancement of technology have been in synch since the industrial revolution. As technology advances, so does the complexity of formulation of these resources.

Current technological trends require a great deal of sophisticated knowledge, both hardware and software supported. This chapter discusses a specific notion of this knowledge, namely the advent of complex heuristics of problem solving.

The notion of *evolutionary heuristics* is one which has its roots in common surrounding. Its premise is that co-operative behavior between many *agents* leads to better and somewhat faster utilisation of the provided resources in the objective of finding the optimal solution to the proposed problem. The optimal solution here refers to a solution, not necessarily the best, but one which can be accepted given the constraints.

Agent based heuristics are those which incorporate a multitude of solutions (unique or replicated) which are then processed using some defined operators to yield new solutions which are presumably better then the previous solutions. These solutions in turn form the next *generation* of solutions. This process iterates for a distinct and predefined number of *generations*.

One of the most prominent heuristic in the scope of real domain problems in Differential Evolution (DE) Algorithm proposed by [31]. Real domain problems are those whose values are essentially real numbers, and the entire solution string can have replicated values. Some of the prominent problems are "De Jong" and "Shwafel" problems which are multi−dimensional.

The aim of the research was to ascertain the feasibility of DE to solve a unique class of problems; Permutative Problems. Permutative problems belong to the Nondeterministic Polynomial–Time hard (NP hard) problems. A problem is assigned to such a class if it is solvable in polynomial time by a nondeterministic Turing Machine.

Two different versions of permutative DE are presented. The first is the Discrete Differential Evolution [23, 26, 6] and its superset Enhanced Differential Evolution Algorithm [8, 9].

3.2 Differential Evolution

In order to describe DE, a schematic is given in Fig 3.1.

There are essentially five sections to the code. Section 1 describes the input to the heuristic. D is the size of the problem, $G_{\max}$ is the maximum number of generations, NP is the total number of solutions, F is the scaling factor of the solution and CR is the factor for crossover. F and CR together make the internal tuning parameters for the heuristic.

Section 2 outlines the initialisation of the heuristic. Each solution $x_{i,j,G=0}$ is created randomly between the two bounds $x^{(lo)}$ and $x^{(hi)}$. The parameter j represents the index to the values within the solution and i indexes the solutions within the population. So, to illustrate, $x_{4,2,0}$ represents the second value of the fourth solution at the initial generation.

After initialisation, the population is subjected to repeated iterations in section 3.

Section 4 describes the conversion routines of DE. Initially, three random numbers r_1, r_2, r_3 are selected, unique to each other and to the current indexed solution i in the population in 4.1. Henceforth, a new index j_{rand} is selected in the solution. j_{rand} points to the value being modified in the solution as given in 4.2. In 4.3, two solutions, $x_{j,r_1,G}$ and $x_{j,r_2,G}$ are selected through the index r_1 and r_2 and their values subtracted. This

1. Input : $D, G_{\max}, NP \geq 4, F \in (0, 1+), CR \in [0,1]$, and initial bounds : $x^{(lo)}, x^{(hi)}$.

2. Initialize : $\begin{cases} \forall i \leq NP \wedge \forall j \leq D : x_{i,j,G=0} = x_j^{(lo)} + rand_j[0,1] \bullet \left(x_j^{(hi)} - x_j^{(lo)}\right) \\ i = \{1,2,...,NP\}, j = \{1,2,...,D\}, G = 0, rand_j[0,1] \in [0,1] \end{cases}$

3. While $G < G_{\max}$

$\forall i \leq NP$

4. Mutate and recombine :

4.1 $r_1, r_2, r_3 \in \{1,2,....,NP\}$, randomly selected, except : $r_1 \neq r_2 \neq r_3 \neq i$

4.2 $j_{rand} \in \{1,2,...,D\}$, randomly selected once each i

4.3 $\forall j \leq D, u_{j,i,G+1} = \begin{cases} x_{j,r_3,G} + F \cdot (x_{j,r_1,G} - x_{j,r_2,G}) \\ \text{if } (rand_j[0,1] < CR \vee j = j_{rand}) \\ x_{j,i,G} \quad \text{otherwise} \end{cases}$

5. Select

$x_{i,G+1} = \begin{cases} u_{i,G+1} & \text{if } f(u_{i,G+1}) \leq f(x_{i,G}) \\ x_{i,G} & \text{otherwise} \end{cases}$

$G = G+1$

Fig. 3.1. Canonical Differential Evolution Algorithm

value is then multiplied by F, the predefined scaling factor. This is added to the value indexed by r_3.

However, this solution is not arbitrarily accepted in the solution. A new random number is generated, and if this random number is less than the value of CR, then the new value replaces the old value in the current solution. Once all the values in the solution are obtained, the new solution is vetted for its fitness or value and if this improves on the value of the previous solution, the new solution replaces the previous solution in the population. Hence the competition is only between the new *child* solution and its *parent* solution.

[32] have suggested ten different working strategies. It mainly depends on the problem on hand for which strategy to choose. The strategies vary on the solutions to be perturbed, number of difference solutions considered for perturbation, and finally the type of crossover used. The following are the different strategies being applied.

Strategy 1: DE/best/1/exp: $\quad u_{i,G+1} = x_{best,G} + F \bullet (x_{r_1,G} - x_{r_2,G})$
Strategy 2: DE/rand/1/exp: $\quad u_{i,G+1} = x_{r_1,G} + F \bullet (x_{r_2,G} - x_{r_3,G})$
Strategy 3: DE/rand−best/1/exp: $u_{i,G+1} = x_{i,G} + \lambda \bullet (x_{best,G} - x_{r_1,G})$
$\quad\quad\quad\quad\quad\quad\quad\quad\quad\quad\quad\quad + F \bullet (x_{r_1,G} - x_{r_2,G})$
Strategy 4: DE/best/2/exp: $\quad u_{i,G+1} = x_{best,G} + F \bullet (x_{r_1,G} - x_{r_2,G} - x_{r_3,G} - x_{r_4,G})$
Strategy 5: DE/rand/2/exp: $\quad u_{i,G+1} = x_{5,G} + F \bullet (x_{r_1,G} - x_{r_2,G} - x_{r_3,G} - x_{r_4,G})$
Strategy 6: DE/best/1/bin: $\quad u_{i,G+1} = x_{best,G} + F \bullet (x_{r_1,G} - x_{r_2,G})$
Strategy 7: DE/rand/1/bin: $\quad u_{i,G+1} = x_{r_1,G} + F \bullet (x_{r_2,G} - x_{r_3,G})$
Strategy 8: DE/rand−best/1/bin: $u_{i,G+1} = x_{i,G} + \lambda \bullet (x_{best,G} - x_{r_1,G})$
$\quad\quad\quad\quad\quad\quad\quad\quad\quad\quad\quad\quad + F \bullet (x_{r_1,G} - x_{r_2,G})$
Strategy 9: DE/best/2/bin: $\quad u_{i,G+1} = x_{best,G} + F \bullet (x_{r_1,G} - x_{r_2,G} - x_{r_3,G} - x_{r_4,G})$
Strategy 10: DE/rand/2/bin: $\quad u_{i,G+1} = x_{5,G} + F \bullet (x_{r_1,G} - x_{r_2,G} - x_{r_3,G} - x_{r_4,G})$

The convention shown is DE/x/y/z. DE stands for Differential Evolution, x represents a string denoting the solution to be perturbed, y is the number of difference solutions considered for perturbation of x, and z is the type of crossover being used (exp: exponential; bin: binomial).

DE has two main phases of crossover: binomial and exponential. Generally a child solution $u_{i,G+1}$ is either taken from the parent solution $x_{i,G}$ or from a mutated donor solution $v_{i,G+1}$ as shown: $u_{j,i,G+1} = v_{j,i,G+1} = x_{j,r_3,G} + F \bullet (x_{j,r_1,G} - x_{j,r_2,G})$.

The frequency with which the donor solution $v_{i,G+1}$ is chosen over the parent solution $x_{i,G}$ as the source of the child solution is controlled by both phases of crossover. This is achieved through a user defined constant, crossover CR which is held constant throughout the execution of the heuristic.

The *binomial* scheme takes parameters from the donor solution every time that the generated random number is less than the CR as given by $rand_j[0,1] < CR$, else all parameters come from the parent solution $x_{i,G}$.

The *exponential* scheme takes the child solutions from $x_{i,G}$ until the first time that the random number is greater than CR, as given by $rand_j[0,1] < CR$, otherwise the parameters comes from the parent solution $x_{i,G}$.

To ensure that each child solution differs from the parent solution, both the exponential and binomial schemes take at least one value from the mutated donor solution $v_{i,G+1}$.

3.2.1 Tuning Parameters

Outlining an absolute value for *CR* is difficult. It is largely problem dependent. However a few guidelines have been laid down [31]. When using binomial scheme, intermediate values of *CR* produce good results. If the objective function is known to be separable, then *CR* = 0 in conjunction with binomial scheme is recommended. The recommended value of *CR* should be close to or equal to 1, since the possibility or crossover occurring is high. The higher the value of *CR*, the greater the possibility of the random number generated being less than the value of *CR*, and thus initiating the crossover.

The general description of *F* is that it should be at least above 0.5, in order to provide sufficient scaling of the produced value.

The tuning parameters and their guidelines are given in Table 3.1

Table 3.1. Guide to choosing best initial control variables

Control Variables	Lo	Hi	Best?	Comments
F: Scaling Factor	0	1.0+	0.3 – 0.9	$F \geq 0.5$
CR: Crossover probability	0	1	0.8 – 1.0	CR = 0, seperable
				CR = 1, epistatic

3.3 Discrete Differential Evolution

The canonical DE cannot be applied to discrete or permutative problems without modification. The internal crossover and mutation mechanism invariably change any applied value to a real number. This in itself will lead to in-feasible solutions.

The objective then becomes one of transformation, either that of the population or that of the internal crossover/mutation mechanism of DE. For this chapter, it was decided not to modify in any way the operation of DE strategies, but to manipulate the population in such a way as to enable DE to operate unhindered.

Since the solution for the population is permutative, a suitable conversion routine was required in order to change the solution from integer to real and then back to integer after crossover. The population was generated as a permutative string. Two conversions routines were devised, one was Forward transformation and the other Backward transformation for the conversion between integer and real values. This new heuristic was termed Discrete Differential Evolution (DDE) [28].

The basic outline DDE is given below.

1. **Initial Phase**
    a) *Population Generation*: An initial number of discrete trial solutions are generated for the initial population.

2. **Conversion**
   a) *Discrete to Floating Conversion*: This conversion schema transforms the parent solution into the required continuous solution.
   b) *DE Strategy*: The DE strategy transforms the parent solution into the child solution using its inbuilt crossover and mutation schemas.
   c) *Floating to Discrete Conversion*: This conversion schema transforms the continuous child solution into a discrete solution.
3. **Selection**
   a) *Validation*: If the child solution is feasible, then it is evaluated and accepted in the next population, if it improves on the parent solution.

3.3.1 Permutative Population

The first part of the heuristic generates the permutative population. A permutative solution is one, where each value within the solution is unique and systematic. A basic description is given in Equation 3.1.

$$P_G = \{x_{1,G}, x_{2,G}, ..., x_{NP,G}\}, \quad x_{i,G} = x_{j,i,G}$$

$$x_{j,i,G=0} = (\text{int})\left(rand_j[0,1] \bullet \left(x_j^{(hi)} + 1 - x_j^{(lo)}\right) + \left(x_j^{(lo)}\right)\right)$$

$$\text{if } x_{j,i} \notin \{x_{0,i}, x_{1,i}, ..., x_{j-1,i}\}$$

$$i = \{1, 2, 3, ..., NP\}, j = \{1, 2, 3, ..., D\} \tag{3.1}$$

where P_G represents the population, $x_{j,i,G=0}$ represents each solution within the population and $x_j^{(lo)}$ and $x_j^{(hi)}$ represents the bounds. The index i references the solution from 1 to NP, and j which references the values in the solution.

3.3.2 Forward Transformation

The transformation schema represents the most integral part of the code. [23] developed an effective routine for the conversion.

Let a set of integer numbers be represented as in Equation 3.2:

$$x_i \in x_{i,G} \tag{3.2}$$

which belong to solution $x_{j,i,G=0}$. The equivalent continuous value for x_i is given as $1 \bullet 10^2 < 5 \bullet 10^2 \leq 10^2$.

The domain of the variable x_i has length = 5 as shown in $5 \bullet 10^2$. The precision of the value to be generated is set to two decimal places (2 d.p.) as given by the superscript two (2) in 10^2. The range of the variable x_i is between 1 and 10^3. The lower bound is 1 whereas the upper bound of 10^3 was obtained after extensive experimentation. The upper bound 10^3 provides optimal filtering of values which are generated close together [27].

The formulation of the forward transformation is given as:

$$x'_i = -1 + \frac{x_i \bullet f \bullet 5}{10^3 - 1} \tag{3.3}$$

Equation 3.3 when broken down, shows the value x_i multiplied by the length 5 and a scaling factor f. This is then divided by the upper bound minus one (1). The value computed is then decrement by one (1). The value for the scaling factor f was established after extensive experimentation. It was found that when f was set to 100, there was a tight grouping of the value, with the retention of optimal filtration's of values. The subsequent formulation is given as:

$$x'_i = -1 + \frac{x_i \bullet f \bullet 5}{10^3 - 1} = -1 + \frac{x_i \bullet f \bullet 5}{10^3 - 1} \quad (3.4)$$

Illustration:

Take a integer value 15 for example. Applying Equation 3.3, we get:

$$x'_i = -1 + \frac{15 \bullet 500}{999} = 6.50751$$

This value is used in the DE internal representation of the population solution parameters so that mutation and crossover can take place.

3.3.3 Backward Transformation

The reverse operation to forward transformation, backward transformation converts the real value back into integer as given in Equation 3.5 assuming x_i to be the real value obtained from Equation 3.4.

$$\text{int}[x_i] = \frac{(1+x_i) \bullet (10^3 - 1)}{5 \bullet f} = \frac{(1+x_i) \bullet (10^3 - 1)}{500} \quad (3.5)$$

The value x_i is rounded to the nearest integer.

Illustration:

Take a continuous value -0.17. Applying equation Equation 3.5:

$$\text{int}[x_i] = \frac{(1 + -0.17) \bullet (10^3 - 1)}{500} = |3.3367| = 3$$

The obtained value is 3, which is the rounded value after transformation.
These two procedures effectively allow DE to optimise permutative solutions.

3.3.4 Recursive Mutation

Once the solution is obtained after transformation, it is checked for feasibility. Feasibility refers to whether the solutions are within the bounds and unique in the solution.

$$x_{i,G+1} = \begin{cases} u_{i,G+1} & \text{if } \begin{cases} u_{j,i,G+1} \neq \{u_{1,i,G+1},...,u_{j-1,i,G+1}\} \\ x^{(lo)} \leq u_{j,i,G+1} \leq x^{(lo)} \end{cases} \\ x_{i,G} \end{cases} \quad (3.6)$$

$$\text{Input}: D, G_{\max}, NP \geq 4, F \in (0, 1+), CR \in [0, 1], \text{bounds}: x^{(lo)}, x^{(hi)}.$$

$$\text{Initialize}: \begin{cases} \forall i \leq NP \land \forall j \leq D \begin{cases} x_{i,j,G=0} = x_j^{(lo)} + rand_j[0,1] \bullet \left(x_j^{(hi)} - x_j^{(lo)}\right) \\ if x_{j,i} \notin \{x_{0,i}, x_{1,i}, ..., x_{j-1,i}\} \end{cases} \\ i = \{1, 2, ..., NP\}, j = \{1, 2, ..., D\}, G = 0, rand_j[0,1] \in [0,1] \end{cases}$$

$$\text{Cost}: \forall i \leq NP : f(x_{i,G=0})$$

$$\begin{cases} \text{While} \quad G < G_{\max} \\ \begin{cases} \text{Mutate and recombine}: \\ r_1, r_2, r_3 \in \{1, 2, ..., NP\}, \text{randomly selected, except}: r_1 \neq r_2 \neq r_3 \neq i \\ j_{rand} \in \{1, 2, ..., D\}, \text{randomly selected once each } i \\ \forall i \leq NP \begin{cases} \forall j \leq D, u_{j,i,G+1} = \begin{cases} (\gamma_{j,r_3,G}) \leftarrow (x_{j,r_3,G}) : (\gamma_{j,r_1,G}) \leftarrow (x_{j,r_1,G}) : (\gamma_{j,r_2,G}) \leftarrow (x_{j,r_2,G}) \\ \quad \text{Forward Transformation} \\ \gamma_{j,r_3,G} + F \cdot (\gamma_{j,r_1,G} - \gamma_{j,r_2,G}) \\ \quad \text{if } (rand_j[0,1] < CR \lor j = j_{rand}) \\ (\gamma_{j,i,G}) \leftarrow (x_{j,i,G}) \text{ otherwise} \end{cases} \\ \left(u'_{i,G+1}\right) = (\rho_{j,i,G+1}) \leftarrow (\varphi_{j,i,G+1}) \quad \text{Backward Transformation} \\ \text{Recursive Mutation}: \\ u_{i,G+1} = \begin{cases} u_{i,G+1} \text{if} \begin{cases} u_{j,i,G+1} \neq \{u_{1,i,G+1}, ..., u_{j-1,i,G+1}\} \\ x^{(lo)} \leq u_{j,i,G+1} \leq x^{(hi)} \end{cases} \\ x_{i,G} \quad \text{otherwise} \end{cases} \\ \text{Select}: \\ x_{i,G+1} = \begin{cases} u_{i,G+1} \text{ if } f(u_{i,G+1}) \leq f(x_{i,G}) \\ x_{i,G} \quad \text{otherwise} \end{cases} \end{cases} \end{cases} \\ G = G + 1 \end{cases}$$

Fig. 3.2. DDE schematic

Recursive mutation refers to the fact that if a solution is deemed in-feasible, it is discarded and the parent solution is retained in the population as given in Equation 3.6.

The general schematic is given in Figure 3.2.

A number of experiments were conducted by DDE on Flowshop Scheduling problems. These are collectively given in the results section of this chapter.

3.4 Enhanced Differential Evolution

Enhanced Differential Evolution (EDE) [7, 8, 9], heuristic is an extension of the DDE variant of DE. One of the major drawbacks of the DDE algorithm was the high frequency of in-feasible solutions, which were created after evaluation. However, since DDE showed much promise, the next logical step was to devise a method, which would repair the in-feasible solutions and hence add viability to the heuristic.

To this effect, three different repairment strategies were developed, each of which used a different index to repair the solution. After repairment, three different enhancement features were added. This was done to add more depth to the code in order to solve permutative problems. The enhancement routines were standard mutation, insertion and local search. The basic outline is given below.

1. **Initial Phase**
    a) *Population Generation*: An initial number of discrete trial solutions are generated for the initial population.

2. **Conversion**
   a) *Discrete to Floating Conversion*: This conversion schema transforms the parent solution into the required continuous solution.
   b) *DE Strategy*: The DE strategy transforms the parent solution into the child solution using its inbuilt crossover and mutation schemas.
   c) *Floating to Discrete Conversion*: This conversion schema transforms the continuous child solution into a discrete solution.
3. **Mutation**
   a) *Relative Mutation Schema*: Formulates the child solution into the discrete solution of unique values.
4. **Improvement Strategy**
   a) *Mutation*: Standard mutation is applied to obtain a better solution.
   b) *Insertion*: Uses a two-point cascade to obtain a better solution.
5. **Local Search**
   a) *Local Search*: 2 Opt local search is used to explore the neighborhood of the solution.

3.4.1 Repairment

In order to repair the solutions, each solution is initially vetted. Vetting requires the resolution of two parameters: firstly to check for any bound offending values, and secondly for repeating values in the solution. If a solution is detected to have violated a bound, it is dragged to the offending boundary.

```
Input : D
Array Solution, ViolateVal, MissingVal
int Counter
for (int i = 0; i < D; i++) {
    for (int j = 0; j < D; j++) {
        if (i == Solution[j]) {
            Counter++; }
    }
    if (Counter > 1) {
        int Index = 0;
        for (int j = 0; j < D; j++) {
            if (i = Solution[j]) {
                Index++
                if (Index > 1) {
                    ViolateVal $\stackrel{Append}{\leftarrow}$ j; }
    }}}
    if (Counter == 0) {
        MissingVal $\stackrel{Append}{\leftarrow}$ i; }
    Counter = 0;
}
```

Fig. 3.3. Pseudocode for replication detection

$$u_{j,i,G+1} = \begin{cases} x^{(lo)} & \text{if } u_{j,i,G+1} < x^{(lo)} \\ x^{(hi)} & \text{if } u_{j,i,G+1} > x^{(hi)} \end{cases} \quad (3.7)$$

Each value, which is replicated, is tagged for its value and index. Only those values, which are deemed replicated, are repaired, and the rest of the values are not manipulated. A second sequence is now calculated for values, which are not present in the solution. It stands to reason that if there are replicated values, then some feasible values are missing. The pseudocode if given in Fig 3.3.

Three unique repairment strategies were developed to repair the replicated values: *front mutation*, *back mutation* and *random mutation*, named after the indexing used for each particular one.

3.4.1.1 Front Mutation

Front mutation indexes the repairment from the front of the replicated array with values randomly selected from the missing value array as shown in Fig 3.4.

```
Array Solution, ViolateVal, MissingVal;
for (int i = 0; i < sizeofViolateVal; i++)
    Solution[ViolateVal[i]] = Random[MissingVal];
}
```

Fig. 3.4. Pseudocode for front mutation

Illustration:

In order to understand *front mutation*, assume an in−feasible solution of dimension $D = 10$: $x = \{3,4,2,1,3,5,6,7,10,5\}$.

The first step is to isolate all repetitive values in the solution. These are highlighted in the following array: $x = \{3,4,2,1,\mathbf{3},\mathbf{5},6,7,10,\mathbf{5}\}$. As shown, the values 3 and 5 are repeated in the solution.

All *first* occurring values are now set as default: $x = \{\mathbf{3},4,2,1,3,\mathbf{5},6,7,10,5\}$. So now only two positions are replicated, index 5 and 10 as given: $x = \{3,4,2,1,3,\underset{5}{5}, 6,7,10,\underset{10}{5}\}$.

An array of missing values is now generated as $MV = \{8,9\}$, since values 8 and 9 are missing from the solution.

An insertion array is now randomly generated, which specifies the position of the insertion of each value: $IA = \{2,1\}$. Since only two values were missing so only two random numbers are generated. In this respect, the first value 2 in *IA*, outlines that the value pointed by index 1 in *MV* which is 8 is to be placed in the second indexed in-feasible solution and likewise for the other missing value given as: $x = \{3,4,2,1,\underset{1}{9},5,6,7,10,\underset{2}{8}\}$.

3.4.1.2 Back Mutation

Back mutation is the opposite of front mutation, and indexes the repairment from the rear of the replicated array as given in Fig 3.5.

> Array $Solution, ViolateVal, MissingVal$;
> for (int $i = sizeof ViolateVal; i > 0; i++$)
> $Solution[ViolateVal[i]] = \text{Random}[MissingVal]$;
> }

Fig. 3.5. Pseudocode for back mutation

Illustration:

In order to understand back mutation assume the same in-feasible solution as in the previous example: $x = \{3,4,2,1,3,5,6,7,10,5\}$.

The first step is to isolate all repetitive values in the solution. These are highlighted in the following array: $x = \{\mathbf{3},4,2,1,\mathbf{3},\mathbf{5},6,7,10,\mathbf{5}\}$. As shown the values 3 and 5 are repeated in the solution.

All last occurring values are now set as default: $x = \{\mathbf{3},4,2,1,3,\mathbf{5},6,7,10,5\}$. So now only two positions are replicated, index 1 and 6 as given: $x = \{\underset{1}{3},4,2,1,3,\underset{6}{5},6, 7,10,5\}$.

An array of missing values is now generated as $MV = \{8,9\}$, since values 8 and 9 are missing from the solution.

An insertion array is now randomly generated, which specifies the position of the insertion of each value: $IA = \{2,1\}$. Since only two values were missing so only two random numbers are generated. In this respect, the first value 1 in IA, outlines that the value pointed by index 1 in MV which is 8 is to be placed in the first indexed in-feasible solution and likewise for the other missing value given as: $x = \{\underset{1}{8},4,2,1,3,\underset{2}{9}, 6,7,10,5\}$.

3.4.1.3 Random Mutation

The most complex repairment schema is the random mutation routine. Each value is selected randomly from the replicated array and replaced randomly from the missing value array as given in Fig 3.6.

> Array $Solution, ViolateVal, MissingVal$;
> for (int $i = sizeof ViolateVal; i > 0; i++$)
> $Solution\left[ViolateVal_{Random[i]}\right] = MissingVal_{Random[i]}$;
> $ViolateVal \overset{delete}{\leftarrow} ViolateVal_{Random[i]}$;
> $MissingVal \overset{delete}{\leftarrow} MissingVal_{Random[i]}$;
> }

Fig. 3.6. Pseudocode for random mutation

Illustration:

Following the previous illustrations, assume the same in-feasible solution: $x = \{3,4,2,1,3,5,6,7,10,5\}$.

The first step is to isolate all repetitive values in the solution. These are highlighted in the following array: $x = \{\mathbf{3},4,2,1,\mathbf{3},\mathbf{5},6,7,10,\mathbf{5}\}$. As shown the values 3 and 5 are repeated in the solution.

A random array is created which sets the default values: $DV = \{2,1\}$, . Here, it shows that the first replicated value which is 3 should be set as default on its second occurrence. The second replicated value 5 should be set as default on its first occurrence: $x = \{\mathbf{3},4,2,1,\mathbf{3},\mathbf{5},6,7,10,\mathbf{5}\}$. The in-feasible values are now in index 1 and 10 given as $x = \{\underset{1}{3},4,2,1,3,5,6,7,10,\underset{10}{5}\}$

An array of missing values is now generated as $MV = \{8,9\}$, since values 8 and 9 are missing from the solution.

An insertion array is now randomly generated, which specifies the position of the insertion of each value: $IA = \{1,2\}$. Since only two values were missing so only two random numbers are generated. In this respect, the first value 1 in IA, outlines that the value pointed by index 1 in MV which is 8 is to be placed in the first indexed in-feasible solution and likewise for the other missing value given as: $x = \{\underset{1}{8},4,2,1,3,5,6,7,10,\underset{10}{9}\}$.

3.4.2 Improvement Strategies

Improvement strategies were included in order to improve the quality of the solutions. Three improvement strategies were embedded into the heuristic. All of these are one time application based. What this entails is that, once a solution is created each strategy is applied only once to that solution. If improvement is shown, then it is accepted as the new solution, else the original solution is accepted in the next population.

3.4.2.1 Standard Mutation

Standard mutation is used as an improvement technique, to explore random regions of space in the hopes of finding a better solution. Standard mutation is simply the exchange of two values in the single solution.

Two unique random values are selected $r_1, r_2 \in rand[1,D]$, where as $r_1 \neq r_2$. The values indexed by these values are exchanged: $Solution_{r_1} \overset{exchange}{\leftrightarrow} Solution_{r_1}$ and the solution is evaluated. If the fitness improves, then the new solution is accepted in the population.

Illustration:

In Standard Mutation assume a solution given as: $x = \{8,4,2,1,3,5,6,7,10,9\}$. Two random numbers are generated within the bounds: $Rnd = \{3,8\}$. These are the indexes of the values in the solution: $x = \{8,4,\underset{3}{2},1,3,5,6,\underset{8}{7},10,9\}$. The values are exchanged $x = \{8,4,7,1,3,5,6,2,10,9\}$ and the solution is evaluated for its fitness.

3.4.2.2 Insertion

Insertion is a more complicated form of mutation. However, insertion is seen as providing greater diversity to the solution than standard mutation.

As with standard mutation, two unique random numbers are selected $r_1, r_2 \in rand$ $[1, D]$. The value indexed by the lower random number $Solution_{r_1}$ is removed and the solution from that value to the value indexed by the other random number is shifted one index down. The removed value is then inserted in the vacant slot of the higher indexed value $Solution_{r_2}$ as given in Fig 3.7.

```
temp = Solution_{r_1};
for (int i = r_1; i < r_2; i++)
    Solution_i = Solution_{i++};
}
Solution_{r_2} = temp;
```

Fig. 3.7. Pseudocode for Insertion

Illustration:

In this Insertion example, assume a solution given as: $x = \{8,4,2,1,3,5,6,7,10,9\}$. Two random numbers are generated within the bounds: $Rnd = \{4,7\}$. These are the indexes of the values in the solution: $x = \{8,4,7,\underset{4}{1},3,5,\underset{7}{6},2,10,9\}$. The lower indexed value is removed from the solution $x = \{8,4,7,\underset{4}{\ },3,5,\underset{7}{6},2,10,9\}$, and all values from the upper index are moved one position down $x = \{8,4,7,|3,5,6,|,2,10,9\}$. The lower indexed value is then slotted in the upper index: $x = \{8,4,7,3,5,6,1,2,10,9\}$.

3.4.3 Local Search

There is always a possibility of stagnation in Evolutionary Algorithms. DE is no exemption to this phenomenon.

Stagnation is the state where there is no improvement in the populations over a period of generations. The solution is unable to find new search space in order to find global optimal solutions. The length of stagnation is not usually defined. Sometimes a period of twenty generation does not constitute stagnation. Also care has to be taken as not be confuse the local optimal solution with stagnation. Sometimes better search space simply does not exist. In EDE, a period of five generations of non-improving optimal solution is classified as stagnation. Five generations is taken in light of the fact that EDE usually operates on an average of hundred generations. This yields to the maximum of twenty stagnations within one run of the heuristic.

$$\alpha = \emptyset$$
$$\text{while} \quad \alpha| < D$$
$$i = rand[1,D], i \notin \alpha$$
$$\beta = \{i\}$$
$$\text{while} \quad \beta| < D$$
$$j = rand[1,D], j \notin \beta$$
$$\text{If} \quad \Delta(x,i,j) < 0; \begin{cases} x_i = x_j \\ x_j = x_i \end{cases}$$
$$\beta = \beta \cup \{j\}$$
$$\alpha = \alpha \cup \{j\}$$

Fig. 3.8. Pseudocode for 2 Opt Local Search

To move away from the point of stagnation, a feasible operation is a neighborhood or local search, which can be applied to a solution to find better feasible solution in the local neighborhood. Local search in an improvement strategy. It is usually independent of the search heuristic, and considered as a plug-in to the main heuristic. The point of note is that local search is very expensive in terms of time and memory. Local search can sometimes be considered as a brute force method of exploring the search space. These constraints make the insertion and the operation of local search very delicate to implement. The route that EDE has adapted is to check the optimal solution in the population for stagnation, instead of the whole population. As mentioned earlier five (5) non-improving generations constitute stagnation. The point of insertion of local search is very critical. The local search is inserted at the termination of the improvement module in the EDE heuristic.

Local Search is an approximation algorithm or heuristic. Local Search works on a *neighborhood*. A complete *neighborhood* of a solution is defined as the set of all solutions that can be arrived at by a move. The word solution should be explicitly defined to reflect the problem being solved. This variant of the local search routine is described in [24] as is generally known as a 2-opt local search.

The basic outline of a Local Search technique is given in Fig 3.8. A number α is chosen equal to zero (0) ($\alpha = \emptyset$). This number iterates through the entire population, by choosing each progressive value from the solution. On each iteration of α, a random number i is chosen which is between the lower (1) and upper (D) bound. A second number β starts at the position i, and iterates till the end of the solution. In this second iteration another random number j is chosen, which is between the lower and upper bound and not equal to value of β. The values in the solution indexed by i and j are swapped. The objective function of the new solution is calculated and only if there is an improvement given as $\Delta(x,i,j) < 0$, then the new solution is accepted.

Illustration:

To understand how this local search operates, consider two solutions x_1 and x_2. The operations parameters of these solutions are:
Upper bound $x^{(hi)} = 5$
Lower bound $x^{(lo)} = 1$
Solution size $D = 5$

Input : $D, G_{\max}, NP \geq 4, F \in (0, 1+), CR \in [0,1]$, and bounds : $x^{(lo)}, x^{(hi)}$.

Initialize : $\begin{cases} \forall i \leq NP \wedge \forall j \leq D \begin{cases} x_{i,j,G=0} = x_j^{(lo)} + rand_j[0,1] \bullet \left(x_j^{(hi)} - x_j^{(lo)}\right) \\ if\ x_{j,i} \notin \{x_{0,i}, x_{1,i}, ..., x_{j-1,i}\} \end{cases} \\ i = \{1, 2, ..., NP\}, j = \{1, 2, ..., D\}, G = 0, rand_j[0,1] \in [0,1] \end{cases}$

Cost : $\forall i \leq NP : f(x_{i,G=0})$

$\begin{cases} \text{While } G < G_{\max} \\ \quad \begin{cases} \text{Mutate and recombine :} \\ r_1, r_2, r_3 \in \{1, 2, ..., NP\}, \text{randomly selected, except :} r_1 \neq r_2 \neq r_3 \neq i \\ j_{rand} \in \{1, 2, ..., D\}, \text{randomly selected once each } i \\ \forall i \leq NP \begin{cases} \forall j \leq D, u_{j,i,G+1} = \begin{cases} (\gamma_{j,r_3,G}) \leftarrow (x_{j,r_3,G}) : (\gamma_{j,r_1,G}) \leftarrow (x_{j,r_1,G}) : \\ (\gamma_{j,r_2,G}) \leftarrow (x_{j,r_2,G}) \quad \text{Forward Transformation} \\ \gamma_{j,r_3,G} + F \cdot (\gamma_{j,r_1,G} - \gamma_{j,r_2,G}) \\ \quad \text{if } (rand_j[0,1] < CR \vee j = j_{rand}) \\ (\gamma_{j,i,G}) \leftarrow (x_{j,i,G}) \quad \text{otherwise} \end{cases} \\ \left(u'_{i,G+1}\right) = \begin{cases} (\rho_{j,i,G+1}) \leftarrow (\varphi_{j,i,G+1}) \text{ Backward Transformation} \\ (u_{j,i,G+1}) \stackrel{mutate}{\leftarrow} (\rho_{j,i,G+1}) \text{ Mutate Schema} \\ \text{if } \left(u'_{j,i,G+1}\right) \notin \{(u_{0,i,G+1}), (u_{1,i,G+1}), ... (u_{j-1,i,G+1})\} \end{cases} \\ (u_{j,i,G+1}) \leftarrow \left(u'_{i,G+1}\right) \text{ Standard Mutation} \\ (u_{j,i,G+1}) \leftarrow \left(u'_{i,G+1}\right) \text{ Insertion} \\ \text{Select :} \\ x_{i,G+1} = \begin{cases} u_{i,G+1} \text{ if } f(u_{i,G+1}) \leq f(x_{i,G}) \\ x_{i,G} \quad \text{otherwise} \end{cases} \end{cases} \end{cases} \\ G = G + 1 \\ \text{Local Search} \quad x_{best} = \Delta(x_{best}, i, j) \quad \text{if stagnation} \end{cases}$

Fig. 3.9. EDE Template

$x_1 = \{2, 5, 4, 3, 1\}$ and $x_1 = \{2, 5, 4, 3, 1\}$
Each value in x_1 and x_2 are paired up and considered.

$$\Delta(i,j) = \begin{Bmatrix} \{2,4\}, \{2,2\}, \{2,1\}, \{2,5\}, \{2,3\}, \\ \{5,4\}, \{5,2\}, \{5,1\}, \{5,5\}, \{5,3\}, \\ \{3,4\}, \{3,2\}, \{3,1\}, \{3,5\}, \{3,3\}, \\ \{1,4\}, \{1,2\}, \{1,1\}, \{1,5\}, \{1,3\} \end{Bmatrix}$$

The cost of the move $\Delta(x,i,j)$ is evaluated. If this value is negative the objective function value for the problem is decrement by $\Delta(x,i,j)$. Hence the solution is improved to a near optimal solution.

The complete template of Enhanced Differential Evolution is given in Fig 3.9.

3.5 Worked Example

This worked example outlines how EDE is used to solve the flowshop scheduling problem. The problem to be solved is the one represented in Table 3.26 (Section 3.6.1: Flow Shop Scheduling Example).

Table 3.2. Table of solutions

Solution	1	2	3	4	5
1	1	2	3	4	5
2	2	1	4	3	5
3	4	5	3	1	2
4	3	1	4	5	2
5	4	2	5	3	1
6	5	4	3	2	1
7	3	5	4	1	2
8	1	2	3	5	4
9	2	5	1	4	3
10	5	3	1	2	4

Table 3.3. Table of initial population with fitness

Fitness	Population				
32	1	2	3	4	5
31	2	1	4	3	5
33	4	5	3	1	2
35	3	1	4	5	2
34	4	2	5	3	1
31	5	4	3	2	1
33	3	5	4	1	2
32	1	2	3	5	4
32	2	5	1	4	3
31	5	3	1	2	4

As presented, this is a 5 job - 4 machine problem.

This example follows the schematic presented in Fig 3.10.

Initially the operating parameters are outlined:

$NP = 10$
$D = 5$
$G_{max} = 1$

For the case of illustration, the operating parameter of NP and G_{max} are kept at a minimum. The other parameters $x^{(lo)}$, $x^{(hi)}$ and D are problem dependent.

Step (1) initialises the population to the required number of solutions.

Since NP is initialised to 10, only 10 permutative solutions are generated. Table 3.2 gives the solution index which represents the positions of each value in the solution in the leading row.

The next procedure is to calculate the objective function of each solution in the population. The time flow matrix for each solution is presented. For detailed explanation on the construction of the time flow matrix, please see Section 3.6.1.

$$Solution_1 = \begin{vmatrix} 6 & 10 & 14 & 19 & 20 \\ 10 & 16 & 18 & 23 & 26 \\ 13 & 19 & 23 & 24 & 29 \\ 17 & 23 & 28 & 31 & 32 \end{vmatrix} \quad Solution_2 = \begin{vmatrix} 4 & 10 & 15 & 19 & 20 \\ 10 & 14 & 19 & 21 & 24 \\ 13 & 17 & 20 & 25 & 28 \\ 17 & 21 & 24 & 30 & 31 \end{vmatrix}$$

$$Solution_3 = \begin{vmatrix} 5 & 6 & 10 & 16 & 20 \\ 9 & 12 & 14 & 20 & 26 \\ 10 & 15 & 19 & 23 & 29 \\ 13 & 26 & 24 & 28 & 33 \end{vmatrix} \quad Solution_4 = \begin{vmatrix} 4 & 10 & 15 & 16 & 20 \\ 6 & 14 & 19 & 22 & 28 \\ 10 & 17 & 20 & 25 & 31 \\ 15 & 21 & 24 & 26 & 35 \end{vmatrix}$$

$$Solution_5 = \begin{vmatrix} 5 & 9 & 10 & 14 & 20 \\ 9 & 15 & 18 & 20 & 24 \\ 10 & 18 & 21 & 25 & 28 \\ 13 & 22 & 23 & 30 & 34 \end{vmatrix} \quad Solution_6 = \begin{vmatrix} 1 & 6 & 10 & 14 & 20 \\ 4 & 10 & 12 & 20 & 24 \\ 7 & 11 & 16 & 23 & 27 \\ 8 & 14 & 21 & 27 & 31 \end{vmatrix}$$

$$Solution_7 = \begin{vmatrix} 4 & 5 & 10 & 16 & 20 \\ 6 & 9 & 14 & 20 & 26 \\ 10 & 13 & 15 & 23 & 29 \\ 15 & 16 & 19 & 27 & 33 \end{vmatrix} \quad Solution_8 = \begin{vmatrix} 6 & 10 & 14 & 15 & 20 \\ 10 & 16 & 18 & 21 & 25 \\ 13 & 19 & 23 & 26 & 27 \\ 17 & 23 & 28 & 29 & 32 \end{vmatrix}$$

$$Solution_9 = \begin{vmatrix} 4 & 5 & 11 & 16 & 20 \\ 10 & 13 & 17 & 21 & 23 \\ 13 & 16 & 20 & 22 & 27 \\ 17 & 18 & 24 & 27 & 32 \end{vmatrix} \quad Solution_{10} = \begin{vmatrix} 1 & 5 & 11 & 15 & 20 \\ 4 & 7 & 15 & 21 & 25 \\ 7 & 11 & 18 & 24 & 26 \\ 8 & 16 & 22 & 28 & 3 \end{vmatrix}$$

The fitness of each solution is given as the last right bottom entry in each solution matrix for that particular solution. The population can now be represented as in Table 3.3.

The optimal value and its corresponding solution, for the current generation is highlighted.

Step (2) is the *forward* transformation of the solution into real numbers. Using Equation 3.3, each value in the solution is transformed. An example of the first $Solution_1 = \{1,2,3,4,5\}$ is given as an illustration:

$$x_1 = -1 + \frac{1 \bullet 500}{999} = -0.499 \quad x_2 = -1 + \frac{2 \bullet 500}{999} = 0.001$$

$$x_3 = -1 + \frac{3 \bullet 500}{999} = 0.501 \quad x_4 = -1 + \frac{4 \bullet 500}{999} = 1.002$$

$$x_5 = -1 + \frac{5 \bullet 500}{999} = 1.502$$

Table 3.4 gives the table with values in real numbers. The results are presented in 3 d.p. format.

In Step (3), DE strategies are applied to the real population in order to find better solutions.

An example of DE operation is shown. Strategy DE/rand/1/exp is used for this example: $u_{i,G+1} = x_{r_1,G} + F \bullet (x_{r_2,G} - x_{r_3,G})$.

3 Forward Backward Transformation

Table 3.4. Table of initial solutions in real number format

Solution Index	1	2	3	4	5
1	0.001	-0.499	1.002	0.501	1.502
2	-0.499	0.001	0.501	1.002	1.502
3	1.002	1.502	0.501	-0.499	0.001
4	0.501	-0.499	1.002	1.502	0.001
5	1.002	0.001	1.502	0.501	-0.499
6	1.502	1.002	0.501	0.001	-0.499
7	0.501	1.502	1.002	-0.499	0.001
8	-0.499	0.001	0.501	1.502	1.002
9	0.001	1.502	-0.499	1.002	0.501
10	1.502	0.501	-0.499	0.001	1.002

Table 3.5. Table of selected solutions

Operation	1	2	3	4	5
X1	-0.499	0.001	0.501	1.002	1.502
X2	0.501	-0.499	1.002	1.502	0.001
X3	1.502	1.002	0.501	0.001	-0.499
(X1 - X2)	-1	0.5	-0.501	-0.5	1.501
F $(X_1 - X_2)$	-0.2	0.1	-0.1002	-0.1	0.3002
X_3 + F $(X_1 - X_2)$	1.302	1.102	0.4008	-0.099	-0.1988

Three random numbers are required to index the solutions in the population given as r_1, r_2 and r_3. These numbers can be chosen as 2, 4 and 6. F is set as 0.2. The procedure is given in Table 3.5.

Table 3.6. Table of final solutions in real number format

Solution Index	1	2	3	4	5
1	-0.435	0.321	0.432	1.543	0.987
2	1.302	1.102	0.401	-0.099	-0.198
3	0.344	1.231	-2.443	-0.401	0.332
4	0.334	-1.043	1.442	0.621	1.551
5	-1.563	1.887	2.522	0.221	-0.432
6	0.221	-0.344	-0.552	0.886	-0.221
7	0.442	1.223	1.423	2.567	0.221
8	-0.244	1.332	0.371	1.421	1.558
9	0.551	0.384	0.397	0.556	0.213
10	-0.532	1.882	-0.345	-0.523	0.512

Table 3.7. Table of solutions with backward transformation

Solution Index	1	2	3	4	5
1	1.128	2.639	2.86	5.08	3.97
2	4.599	4.199	2.798	1.8	1.6
3	2.685	4.457	-2.883	1.196	2.661
4	2.665	-0.085	4.879	3.238	5.096
5	-1.124	5.768	7.036	2.439	1.134
6	2.439	1.31	0.895	3.768	1.556
7	2.881	4.441	4.841	7.126	2.439
8	1.51	4.659	2.739	4.837	5.11
9	3.098	2.765	2.791	3.108	2.423
10	0.935	5.758	1.308	0.953	3.02

Table 3.8. Rounded solutions

Solution Index	1	2	3	4	5
1	1	3	3	5	4
2	5	4	3	2	2
3	3	4	-3	1	3
4	3	-1	5	3	5
5	-1	6	7	2	1
6	2	1	1	4	2
7	3	4	5	7	2
8	2	5	3	5	5
9	3	3	3	3	2
10	1	6	1	1	3

Using the above procedure the final solution for the entire population can be given as in Table 3.6.

Backward transformation is applied to each solution in Step (4). Taking the first $Solution_1 = \{-0.435, 0.321, 0.432, 1.543, 0.987\}$, a illustrative example is given using Equation 3.5.

$$x_1 = \frac{(1+-0.435) \bullet 999}{500} = 1.128 \qquad x_2 = \frac{(1+0.001) \bullet 999}{500} = 2.639$$

$$x_3 = \frac{(1+0.501) \bullet 999}{500} = 2.86 \qquad x_4 = \frac{(1+1.002) \bullet 999}{500} = 5.08$$

$$x_5 = \frac{(1+1.502) \bullet 999}{500} = 3.97$$

The *raw* results are given in Table 3.7 with tolerance of 3 d.p.

Each value in the population is rounded to the nearest integer as given in Table 3.8.

3 Forward Backward Transformation   53

Table 3.9. Bounded solutions

Solution Index	1	2	3	4	5
1	1	3	3	5	4
2	5	4	3	2	2
3	3	4	1	1	3
4	3	1	5	3	5
5	1	5	5	2	1
6	2	1	1	4	2
7	3	4	5	5	2
8	2	5	3	5	5
9	3	3	3	3	2
10	1	5	1	1	3

Table 3.10. Replucated values

Solution Index	1	2	3	4	5
1	1	3	3	5	4
2	5	4	3	2	2
3	3	4	1	1	3
4	3	1	5	3	5
5	**1**	5	5	2	1
6	**2**	1	1	4	2
7	3	4	5	5	2
8	2	5	3	5	5
9	**3**	3	3	3	2
10	**1**	5	**1**	**1**	3

Recursive mutation is applied in Step (5). For this illustration, the random mutation schema is used as this was the most potent and also the most complicated.

The first routine is to drag all bound offending values to the offending boundary. The boundary constraints are given as $x^{(lo)} = 1$ and $x^{(hi)} = 5$ which is lower and upper bound of the problem. Table 3.9 gives the *bounded* solution.

In *random mutation*, initially all the duplicated values are isolated as given in Table 3.10.

The next step is to randomly set default values for each replication. For example, in Solution 1, the value 3 is replicated in 2 indexes; 2 and 3. So a random number is generated to select the default value of 3. Let us assume that index 3 is generated. In this respect, only value 3 indexed by 2 is labelled as replicated. This routine is applied to the entire population, solution piece wise in order to set the default values.

A possible representation can be given as in Table 3.11.

Table 3.11. Ramdomly replaced values

Solution Index	1	2	3	4	5
1	1	**3**	*3*	**5**	**4**
2	5	4	3	**2**	2
3	**3**	4	*1*	**1**	**3**
4	**3**	1	5	*3*	*5*
5	**1**	5	**5**	2	*1*
6	2	*1*	**1**	4	**2**
7	3	4	5	**5**	2
8	2	**5**	3	*5*	**5**
9	**3**	**3**	*3*	**3**	2
10	*1*	5	**1**	**1**	**3**

Table 3.12. Missing values

Solution Index	1	2	3
1	2		
2	1		
3	2	5	
4	2	4	
5	3	4	
6	3	5	
7	1		
8	1	4	
9	1	4	5
10	2	4	

The italicised values in Table 3.11 have been selected as default through randomisation. The next phase is to find those values which are not present in the solution. All the missing values in the solutions are given in Table 3.12.

In the case of Solutions 1, 2 and 7, it is very simple to repair the solution, since there is only one missing value. The missing value is simply placed in the replicated index for that solution. In the other cases, *positional indexes* are randomly generated. A positional index tells as to where the value will be inserted in the solution. A representation is given in Table 3.13.

Table 3.13 shows that the *first* missing value will be placed in the *second* replicated value index in the solution, and the *second* missing value will be placed in the *first* replicated index value. The final placement is given in Table 3.14.

The solutions are now permutative. The fitness for each solution is calculated in Table 3.15.

3 Forward Backward Transformation 55

Table 3.13. Positional Index

Solution Index	1	2	3
1	2		
2	1		
3	2	5	
4	2	4	
5	3	4	
6	3	5	
7	1		
8	1	4	
9	1	4	5
10	2	4	

Table 3.14. Final placement of missing values

Solution Index	1	2	3	4	5
1	1	2	3	5	4
2	5	4	3	2	1
3	5	4	1	2	3
4	2	1	4	3	5
5	3	5	4	2	1
6	2	1	5	4	3
7	3	4	5	1	2
8	2	1	3	5	4
9	1	5	3	4	2
10	1	5	4	2	3

Table 3.15. Fitness of new population

Fitness	Population				
32	1	2	3	5	4
31	5	4	3	2	1
34	5	4	1	2	3
31	2	1	4	3	5
31	3	5	4	2	1
32	2	1	5	4	3
33	3	4	5	1	2
30	2	1	3	5	4
33	1	5	3	4	2
35	1	5	4	2	3

Table 3.16. Random Index

Solution	Index	
1	4	2
2	1	4
3	2	3
4	3	5
5	1	5
6	2	4
7	1	2
8	3	4
9	3	1
10	2	4

Table 3.17. Fitness of new *mutated* population

Solution	Fitness	Solution Index				
		1	2	3	4	5
1	32	1	5	3	2	4
2	31	2	4	3	5	1
3	34	5	1	4	2	3
4	32	2	1	5	3	4
5	35	1	5	4	2	3
6	33	2	4	5	1	3
7	32	4	3	5	1	2
8	32	2	1	5	3	4
9	33	3	5	1	4	2
10	35	1	2	4	5	3

Table 3.18. Population after *mutation*

Fitness	Population				
32	1	2	3	5	4
31	5	4	3	2	1
34	5	4	1	2	3
31	2	1	4	3	5
31	3	5	4	2	1
32	2	1	5	4	3
32	4	3	5	1	2
30	2	1	3	5	4
33	1	5	3	4	2
35	1	5	4	2	3

3 Forward Backward Transformation 57

Table 3.19. Random Index

Solution	Index	
1	2	4
2	1	3
3	2	5
4	1	5
5	2	4
6	3	5
7	1	4
8	3	5
9	1	4
10	2	5

Table 3.20. Insertion process 1

Index	1	2	3	4	5
Solution 1	1		3	5	4

Table 3.21. Insertion process 2

Index	1	2	3	4	5
Solution 1	1	3	5		4

Step (6) describes the *Standard Mutation* schema. In standard mutation, a single value swap occurs. Assume that a list of random indexes in Table 3.16 are generated which show which values are to be swapped.

It can be seen from Table 3.16, that the values indexed by 4 and 2 are to be swapped in Solution 1 and so forth for all the other solutions. The new *possible* solutions are given in Table 3.17 with their calculated fitness values. The highlighted values are the mutated values.

Only solution 7 improved in the *mutation schema* and replaces the old solution on position 7 in the population. The final population is given in Table 3.18.

Step (7), *Insertion* also requires the generation of random indexes for cascading of the solutions. A new set of random numbers can be visualized as in Table 3.19.

In Table 3.19 the values are presented in ascending order. Taking solution 1, the first process is to remove the value indexed by the first lower index (2) as shown in Table 3.20.

The second process is to move all the values from the upper index (4) to the lower index as in Table 3.21.

The last part is to insert the first removed value from the lower index into the place of the now vacant upper index aas shown in Table 3.22.

Table 3.22. Insertion process 3

Index	1	2	3	4	5
Solution 1	1	3	5	2	4

Table 3.23. Population after *insertion*

Fitness			Population		
31	**1**	**3**	**5**	**2**	**4**
31	4	3	5	2	1
32	**5**	**1**	**2**	**3**	**4**
33	1	4	3	5	2
33	3	4	2	5	1
31	**2**	**1**	**4**	**3**	**5**
33	3	5	1	4	2
32	2	1	5	4	3
33	5	3	4	1	2
34	**1**	**4**	**2**	**3**	**5**

Table 3.24. Final population

Fitness			Population		
31	1	3	5	2	4
31	4	3	5	2	1
32	5	1	2	3	4
33	1	4	3	5	2
33	3	4	2	5	1
31	2	1	4	3	5
33	3	5	1	4	2
32	2	1	5	4	3
33	5	3	4	1	2
34	1	4	2	3	5

Likewise, all the solutions are *cascaded* in the population and their new fitness calculated. The population is then represented as in Table 3.23.

After *Insertion*, four better solutions were found. These solutions replace the older solution in the population. The final population is given in Table 3.24.

DE postulates that each *child* solution replaces it direct *parent* in the population if it has better fitness. Comparing the final population in Table 3.24 with the initial population in Table 3.2, it can be seen that seven solutions produced even or better fitness than the solutions in the old population. Thus these child solutions replace the parent solutions in the population for the next generation as given in Table 3.25.

Table 3.25. Final population with fitness

Fitness	Population				
31	1	3	5	2	4
31	5	4	3	2	1
33	4	5	3	1	2
31	2	1	4	3	5
31	3	5	4	2	1
31	2	1	4	3	5
32	4	3	5	1	2
30	2	1	3	5	4
32	2	5	1	4	3
31	5	3	1	2	4

The new solution has a fitness of 30, which is a new fitness from the previous generation. This population is then taken into the next generation. Since we specified the $G_{max} = 1$, only 1 iteration of the routine will take place.

Using the above outlined process, it is possible to formulate the basis for most permutative problems.

3.6 Flow Shop Scheduling

One of the common manufacturing tasks is *scheduling*. Often in most manufacturing systems, a number of tasks have to be completed on every *job*. Usually all these jobs have to follow the same route through the different *machines*, which are set up in a series. Such an environment is called a *flow shop* (FSS) [30].

The standard three-field notation [20] used is that for representing a scheduling problem as $\alpha|\beta|F(C)$, where α describes the machine environment, β describes the deviations from standard scheduling assumptions, and $F(C)$ describes the objective C being optimised. This research solves the generic flow shop problem represented as $n/m/F||F(C_{max})$.

Stating these problem descriptions more elaborately, the minimization of completion time (makespan) for a flow shop schedule is equivalent to minimizing the objective function $\Im$:

$$\Im = \sum_{j=1}^{n} C_{m,j} \tag{3.8}$$

s.t.

$$C_{i,j} = \max\left(C_{i-1,j}, C_{i,j-1}\right) + P_{i,j} \tag{3.9}$$

where, $C_{m,j}$ = the completion time of job j, $C_{i,j} = k$ (any given value), $C_{i,j} = \sum_{k=1}^{j} C_{1,k}$; $C_{i,j} = \sum_{k=1}^{j} C_{k,1}$ machine number, j job in sequence, $P_{i,j}$ processing time of job j on

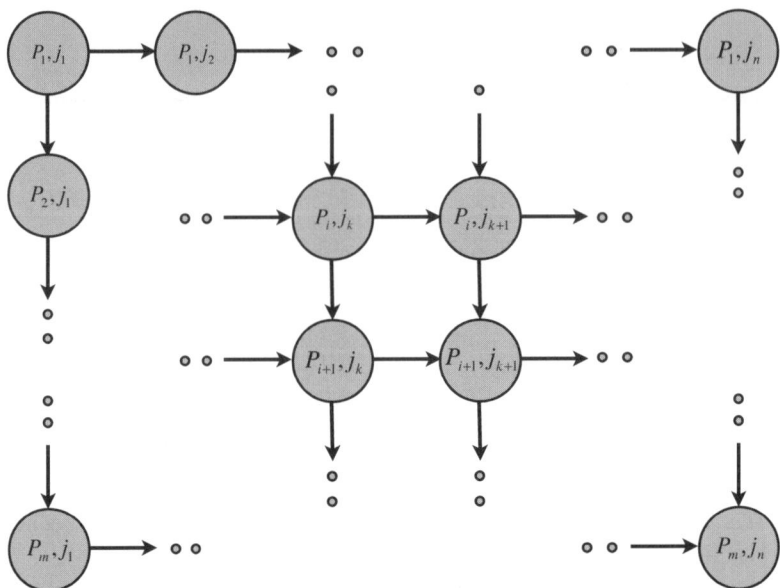

Fig. 3.10. Directed graph representation for the makespan

machine i. For a given sequence, the mean flow time, $MFT = \frac{1}{n} \sum_{i=1}^{m} \sum_{j=1}^{n} c_{ij}$, while the condition for tardiness is $c_{m,j} > d_j$. The constraint of Equation 3.9 applies to these two problem descriptions.

3.6.1 Flow Shop Scheduling Example

Generally, two versions of flowshop problems exist. Finding an optimal solution when the sequence changes within the schedule are flexible and changes allowed are generally harder to formulate and calculate. The schedules which are fixed are simpler to calculate and are known as permutative flow shops.

A simple representation of flowshop is given through the *directed graph method*. The *critical path* in the *directed graph* gives the makespan for the current schedule. For a given sequence $j_1,...,j_n$, the graph is constructed as follows: For each operation of a specific job j_k on a specific machine i, there is a node (i, j_k) with the *processing time* for that job on that machine. Node (i, j_k), $i = 1,...,m-1$ and $k = 1,....,n-1$, has arcs going to nodes $(i+1, j_k)$ and (i, j_{k+1}). Nodes corresponding to machine m have only one outgoing arc, as do the nodes in job j_n. Node (m, j_n), has no outgoing arcs as it is the terminating node and the total weight of the path from first to last node is the makespan for that particular schedule [30]. A schmetic is given in Fig 3.10.

Assume a representation of five jobs on four machines given in Table 3.26.

Given a schedule $\{1,2,3,4,5\}$ which is the schedule $\{j_1, j_2, j_3, j_4, j_5\}$, implying that all jobs in that sequence will transverse all the machines.

Table 3.26. Example of job times

jobs	j_1	j_2	j_3	j_4	j_5
P_{1,j_k}	6	4	4	5	1
P_{2,j_k}	4	6	2	4	3
P_{3,j_k}	3	3	4	1	3
P_{4,j_k}	4	4	5	3	1

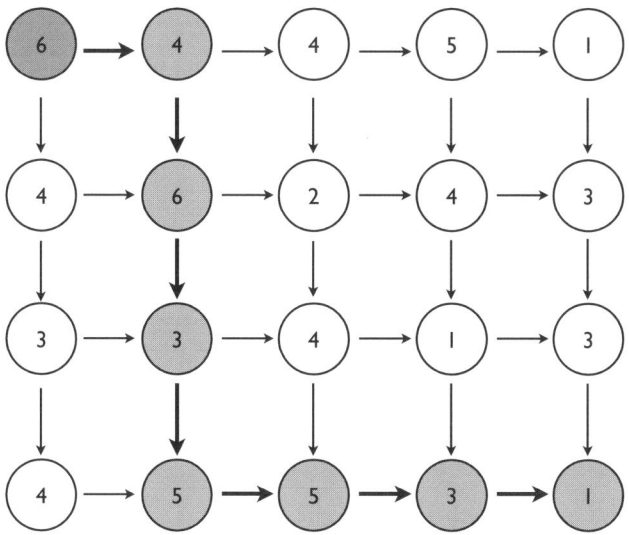

Fig. 3.11. Directed graph representation of the schedule

The directed graph representation for this schedule is given in Fig 3.11.

Each node on the graph represents the time taken to process that particular job on that particular machine. The bold lines represent the *critical path* for that particular schedule.

The Gantt chart for this schedule is represented in Fig 3.12.

The critical path is highlighted The critical path represents jobs, which are not delayed or buffered. This is important for those shops, which have machines with no buffering between them. The total time for this schedule is 34. However, from this representation, it is difficult to make out the time. A better representation of the directed graph and critical path is given in Fig. 3.13.

The cumulative time nodes gives the time accumulated at each node. The final node gives the makespan for the total schedule.

The total time Gantt chart is presented in Fig 3.14.

As the schedule is changed, so does the directed graph.

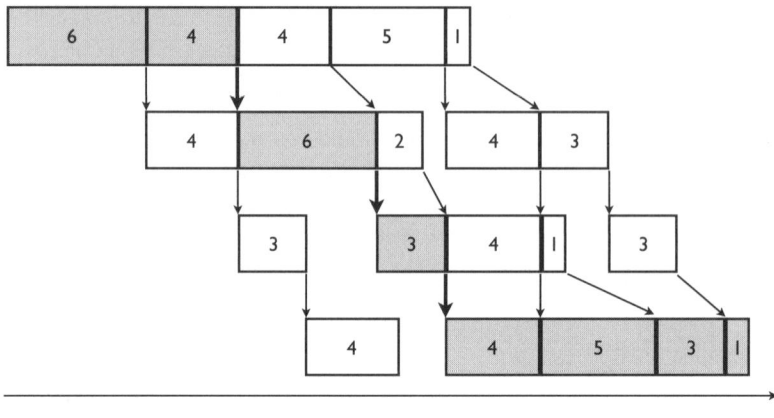

Fig. 3.12. Gantt chart representation of the schedule

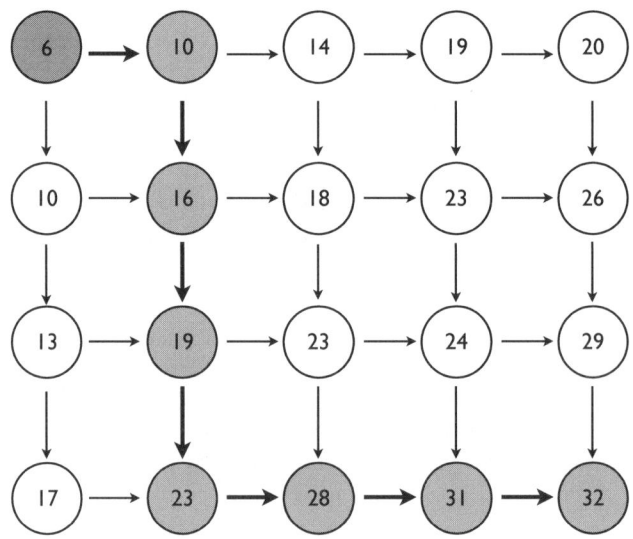

Fig. 3.13. Directed time graph and critical path

3.6.2 Experimentation for Discrete Differential Evolution Algorithm

The first phase of experimentation was used on FSS utilising DDE algorithm. Eight varying problem instances were selected from the literature, which represents a range of problem complexity. The syntax of the problem *n x m* represents *n* machines and *m* jobs. These problem instances were generated randomly for previous tests and range from small problem types (*4x4* to *15x25*), medium problem type (*20x50*) and large problem types (*25x75* and *30x100*).

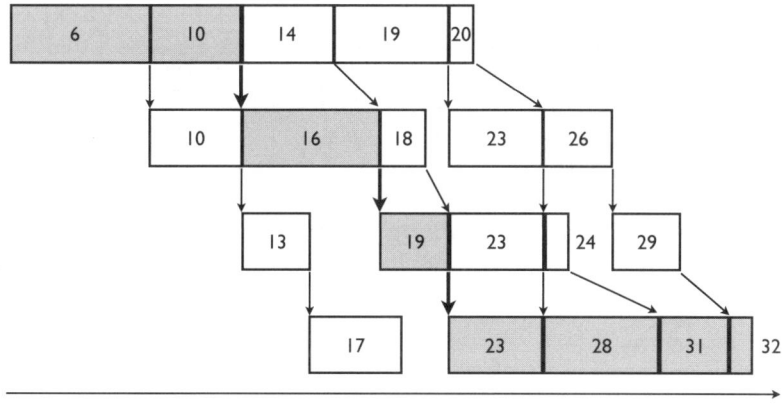

Fig. 3.14. Accumulated time Gantt Chart

Table 3.27. DDE FSS operational values

Parameter	Values
NP	150
CR	0.9
F	0.3

Table 3.28. Comparison of 10 DE-strategies using the *10x25* problem data set

	Strategy									
	1	2	3	4	5	6	7[a]	8	9	10
Makespan	211.8	209.2	212.2	212.4	208.6	210.6	207.8	212.4	210	207.2
Total tardiness	3001.8	3034.6	3021.4	3089.2	3008	2987.8	2936.4	3034.2	2982.8	2990.6
Mean flowtime	105.75	105.11	105.52	107.71	104.68	103.03	103.17	105.32	104.7	104.16

[a]Strategy 7 is the best.

In order to operate, the first phase is to obtain the optimal tuning parameters. All parameters were obtained empirically. The values are given in Table 3.27.

The second phase was to obtain the optimal strategy. Through experience in solving these problems, it became evidently clear that not all the strategies behaved similarly, hence the need to isolate the most promising one from the ten different.

An arbitrary problem of average difficulty was selected, in this case the *10x25* job problem, and using the selected parameters, ten iterations were done. The average values are presented in Table 3.28. Using the multi-objective function of makespan, tardiness and flowtime, Strategy 7 was selected as the optimal.

Table 3.29. DDE FSS makespan

m x n	Generated problems	GA	DE	(Solution) GA/DE
4x4	5	44	39	-
5x10	5	79	79	-
8x15	5	143	138	-
10x25	5	205	202	-
15x25	5	248	253	98.02
20x50	5	468	470	99.57
25x75	5	673	715.4	94.07
30x100	5	861	900.4	95.62

Table 3.30. DDE FSS total tardiness

m x n	Generated problems	GA	DE	(Solution) GA/DE
4x4	5	54	52.6	-
5x10	5	285	307	92.83
8x15	5	1072	1146	93.54
10x25	5	2869	2957	97.02
15x25	5	3726	3839.4	97.06
20x50	5	13683	14673.6	93.25
25x75	5	30225	33335.6	90.67
30x100	5	51877	55735.6	93.07

With all the experimentation parameters selected, the FSS problems were evaluated. Three different objective functions were to be analysed. The first was the makespan. The makespan is equivalent to the completion time for the last job to leave the system. The results are presented in Table 3.29.

The second objective is the tardiness. Tardiness relates to the number of tardy jobs; jobs which will not meet their due dates and which are scheduled last. This reflects the on-time delivery of jobs and is of paramount importance to production planning and control [30]. The results are given in Table 3.30.

The final objective is the mean flowtime of the system. It is the sum of the weighted completion time of the n jobs which gives an indication of the total holding or inventory costs incurred by the schedule. The results are presented in Table 3.31.

Tables 3.29 − 3.31 show the comparison between Genetic Algorithm (GA) developed in a previous study for flowshop scheduling [28], compared with DDE. Upon analysis it is seen that, DE algorithm performs better than GA for small-sized problems, and competes appreciably with GA for medium to large-sized problems. These results are not compared to the traditional methods since earlier study of [4] show that GA based algorithm for flow shop problems outperform the best existing traditional approaches such as the ones proposed by [16] and [39].

Table 3.31. DDE Mean Flowtime

m x n	Generated problems	GA	DE	(Solution) GA/DE
4x4	5	21.38	22.11	-
5x10	5	35.3	36.34	97.14
8x15	5	63.09	66.41	95
10x25	5	98.74	103.89	95.04
15x25	5	113.85	122.59	93.03
20x50	5	216	234.32	92.18
25x75	5	317	354.77	89.35
30x100	5	399.13	435.49	91.56

Table 3.32. EDE FSS operational values

Parameter	Values
Strategy	9
NP	150
CR	0.3
F	0.1

These obtained results formed the basic for the enhancement of DDE. It should be noted that even with a very high percentage of in-feasible solutions obtained, DDE managed to outperform GA.

3.6.3 Experimentation for Enhanced Differential Evolution Algorithm

The second phase of experiments outline experimentation of EDE to FSS. As with the DDE, operational parameters were empirically obtained as given in Table 3.32. As can be noticed the parameters are very different from those used in DDE for the same problems. This is attributed to the new routines added to DDE which adds another layer of stochastically to EDE.

The first section of experimentation was conducted on the same group of FSS problems as GA and DDE to obtain comparison results. In this respect, only makespan was evaluated. For all the problem instances, EDE performs optimally compared to the other two heuristics. Columns 5 to 7 in Table 3.33 gives the effectiveness comparisons of EDE, DDE and GA, with EDE outperforming both DDE and GA.

With the validation completed for EDE, more extensive experimentation was conducted to test its complete operational range in FSS.

The second set of benchmark problems is from the three papers of [3], [33] and [15]. All these problem sets are available in the OR Library [29]. The EDE results are compared with the optimal values reported for these problems as given in Table 3.34. The conversion is given in Equation 3.10:

$$\Delta = \frac{(H-U) \bullet 100}{U} \qquad (3.10)$$

where H represents the obtained value and U is the reported optimal. For the Car and Hel set of problems, EDE easily obtains the optimal values, and on average around 1% above the optimal for the reC instances.

Table 3.33. FSS comparison

	DDE	GA	EDE	% DDE–GA	% EDE–DDE	% EDE–GA
F 5 x 10	79.4	-	78	-	101.79	-
F 8 x 15	138.6	143	134	103.17	103.43	106.71
F 10 x 25	207.6	205	194	98.74	107.01	105.67
F 15 x 25	257.6	248	240	96.27	107.33	103.33
F 20 x 50	474.8	468	433	98.56	109.65	108.08
F 25 x 75	715.4	673	647	94.07	110.57	104.01
F 30 x 100	900.4	861	809	95.62	111.29	106.42
Ho Chang	213	213	213	100	100	100

Table 3.34. Comparison of FSS instances

Instance	Size	Optimal	EDE	% to Optimal
Car 1	11 x 5	7038	7038	0
Car 2	13 x 4	7166	7166	0
Car 3	12 x 5	7312	7312	0
Car 4	14 x 4	8003	8003	0
Car 5	10 x 6	7720	7720	0
Car 6	8 x 9	8505	8505	0
Car 7	7 x 7	6590	6590	0
Car 8	8 x 8	8366	8366	0
Hel 2	20 x 10	135	135	0
reC 01	20 x 5	1247	1249	0.16
reC 03	20 x 5	1109	1111	0.18
reC 05	20 x 5	1242	1249	0.56
reC 07	20 x 10	1566	1584	1.14
reC 09	20 x 10	1537	1574	2.4
reC 11	20 x 10	1431	1464	2.3
reC 13	20 x 15	1930	1957	1.39
reC 15	20 x 15	1950	1984	1.74
reC 17	20 x 15	1902	1957	2.89
reC 19	30 x 10	2093	2132	1.86
reC 21	30 x 10	2017	2065	2.37
reC 23	30 x 10	2011	2073	3.08

Table 3.35. EDE comparison with DE_{spv} and PSO over the Taillard benchmark problem

	GA		PSO_{spv}		DE_{spv}		$DE_{spv+exchange}$		EDE	
	Δ_{avg}	Δ_{std}	Δ_{avg}	Δ_{std}	Δ_{avg}	Δ_{std}	Δ_{avg}	Δ_{std}	Δ_{avg}	Δ_{std}
20x5	3.13	1.86	1.71	1.25	2.25	1.37	0.69	0.64	0.98	0.66
20x10	5.42	1.72	3.28	1.19	3.71	1.24	2.01	0.93	1.81	0.77
20x20	4.22	1.31	2.84	1.15	3.03	0.98	1.85	0.87	1.75	0.57
50x5	1.69	0.79	1.15	0.7	0.88	0.52	0.41	0.37	0.4	0.36
50x10	5.61	1.41	4.83	1.16	4.12	1.1	2.41	0.9	3.18	0.94
50x20	6.95	1.09	6.68	1.35	5.56	1.22	3.59	0.78	4.05	0.65
100x5	0.81	0.39	0.59	0.34	0.44	0.29	0.21	0.21	0.41	0.29
100x10	3.12	0.95	3.26	1.04	2.28	0.75	1.41	0.57	1.46	0.36
100x20	6.32	0.89	7.19	0.99	6.78	1.12	3.11	0.55	3.61	0.36
200x10	2.08	0.45	2.47	0.71	1.88	0.69	1.06	0.35	0.95	0.18

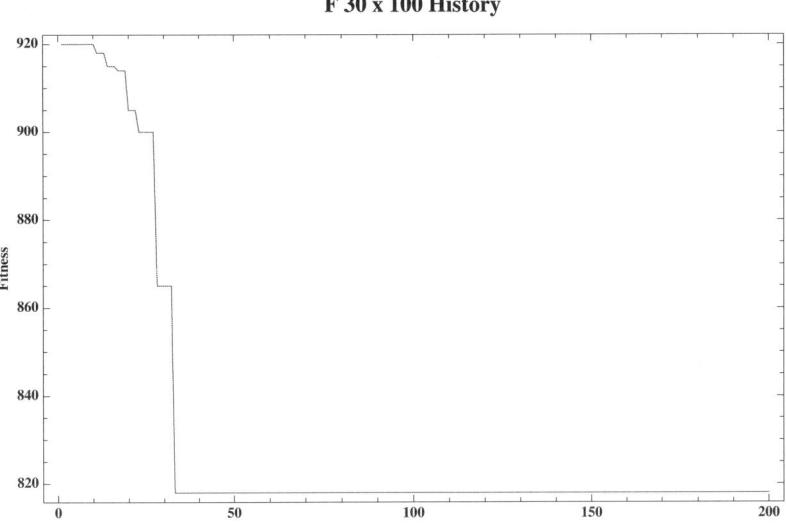

Fig. 3.15. Sample output of the F30x100 FSS problem.

The third experimentation module is referenced from [37]. These sets of problems have been extensively evaluated (see [22, 34]). This benchmark set contains 100 particularly hard instances of 10 different sizes, selected from a large number of randomly generated problems.

A maximum of ten iterations was done for each problem instance. The population was kept at 100, and 100 generations were specified. The results represented in Table 3.35, are as quality solutions with the percentage relative increase in makespan with respect to the upper bound provided by [37] as given by Equation 3.10.

The results obtained are compared with those produced by GA, Particle Swarm Optimisation (PSO_{spv}) DE (DE_{spv}) and DE with local search ($DE_{spv+exchange}$) as in [38]. The results are tabulated in Table 3.35.

It can be observed that EDE compares outstandingly with other algorithms. EDE basically outperforms GA, PSO and DE_{spv}. The only serious competition comes from the new variant of $DE_{spv+exchange}$. EDE and $DE_{spv+exchange}$ are highly compatible. EDE outperforms $DE_{spv+exchange}$ on the data sets of 20x10, 20x20, 50x5 and 200x5. In the remainder of the sets EDE performs remarkbly to the values reported by $DE_{spv+exchange}$. On average EDE displays better standard deviation than that of $DE_{spv+exchange}$. This validates the consistency of EDE compared to $DE_{spv+exchange}$. It should be noted that $DE_{spv+exchange}$ utilises local search routine as its search engine.

A sample generation for the *F 30 x 100* FSS problem is given in Fig 3.15.

3.7 Quadratic Assignment Problem

The second class of problems to be conducted by EDE was the *Quadratic Assignment Problem* (QAP). QAP is a *NP*-hard optimisation problem [35] which was stated for the first time by [18]. It is considered as one of the hardest optimisation problems as general instances of size $n \geq 20$ cannot be solved to optimally [10].

It can be described as follows: Given two matrices

$$A = (a_{ij}) \tag{3.11}$$

$$B = (b_{ij}) \tag{3.12}$$

find the permutation π^* minimising

$$\min_{\pi \in \Pi(n)} f(\pi) = \sum_{i=1}^{n} \sum_{j=1}^{n} a_{ij} \bullet b_{\pi(i)\pi(j)} \tag{3.13}$$

where $\prod(n)$ is a set of permutations of *n* elements.

The problem instances selected for the QAP are from the OR Library [29] and reported in [13]. There are two separate problem modules; *regular* and *irregular*.

The difference between regular and irregular problems is based on the *flow–dominance*. Irregular problems have a flow–dominance statistics larger than 1.2. Most of the problems come from practical applications or have been randomly generated with non-uniform laws, imitating the distributions observed in real world problems.

In order to differentiate among the classes of QAP instances, the flow dominance *fd* is used. It is defined as a coefficient of variation of the flow matrix entries multiplied by 100. That is:

$$fd = \frac{100\sigma}{\mu} \tag{3.14}$$

where:

$$\mu = \frac{1}{n^2} \bullet \sum_{i=1}^{n} \sum_{j=1}^{n} b_{ij} \tag{3.15}$$

$$\sigma = \sqrt{\frac{1}{n^2} \cdot \sum_{i=1}^{n} \sum_{j=1}^{n} (b_{ij} - \mu)^2} \qquad (3.16)$$

3.7.1 Quadratic Assignment Problem Example

As example for the QAP is given as the faculty location problem given in Fig 3.16.

The objective is to allocate location to faculties. There is a specific distance between the faculties, and there is a specified flow between the different faculties, as shown by the thickness of the lines. An arbitrary schedule can be $\{2,1,4,3\}$, as given in Fig 3.16. Two distinct matrices are required: one *distance* and one *flow* matrix as given in
Tables 3.36 and 3.37.

Applying the QAP formula, the function becomes:

$$Sequence = \begin{Bmatrix} D(1,2) \bullet F(1,2) + \\ D(1,3) \bullet F(2,4) + \\ D(2,3) \bullet F(1,4) + \\ D(3,4) \bullet F(3,4) \end{Bmatrix}$$

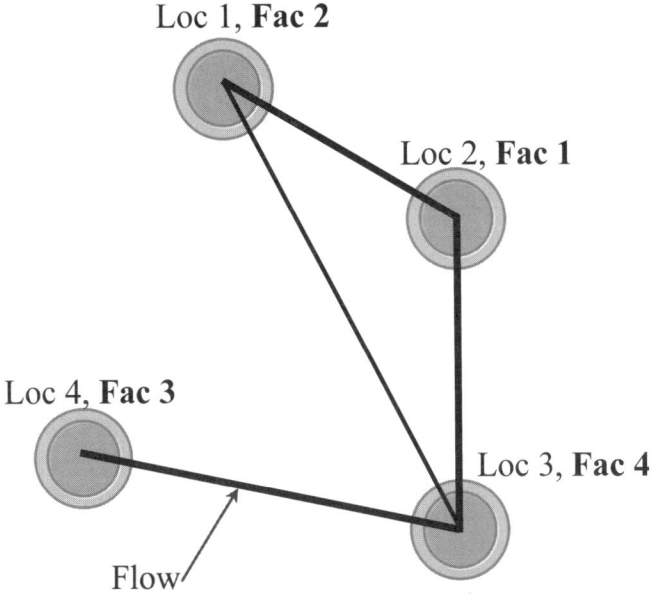

Fig. 3.16. Faculty location diagram for 2, 1, 4, 3

Table 3.36. Distance Matrix

Distance	Value
D(1,2)	22
D(1,3)	53
D(2,3)	40
D(3,4)	55

Table 3.37. Flow Matrix

Flow	Value
F(2,4)	1
F(1,4)	2
F(1,2)	3
F(3,4)	4

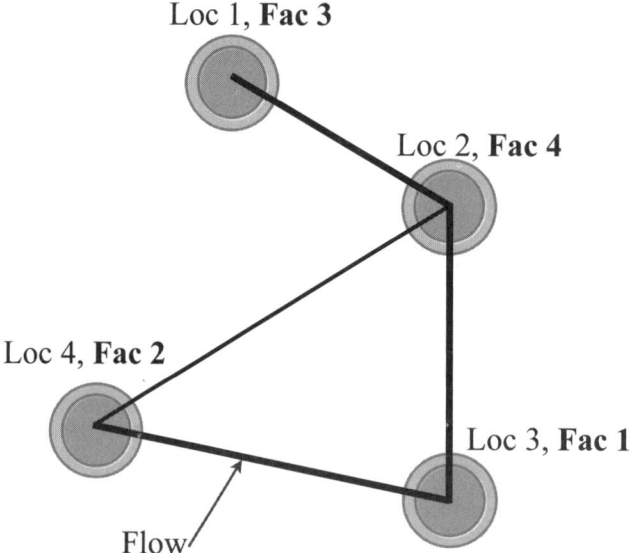

Fig. 3.17. Faculty location diagram for 3, 4, 1, 2

$$Cost = \begin{Bmatrix} (22 \bullet 3) + \\ (53 \bullet 1) + \\ (40 \bullet 2) + \\ (55 \bullet 4) \end{Bmatrix} = 419$$

Now, assume a different permutation: $\{3,4,1,2\}$. The faculty location diagram is now given in Fig 3.17.

The solution for this permutation is 395, The flow matrix remains the same, and only the distance matrix changes to reflect the new faculty location.

3.7.2 Experimentation for Irregular QAP

The first phase as with FSS, was to empirically obtain the operational values as given in Table 3.38. These values were used for both regular and irregular instances.

The first set of experimentations was on irregular instances. These are those with flow dominance of greater than 1.

The results are presented in Table 3.39. The results are presented as the factor distance from the optimal: $\Delta = \frac{(H-U)}{U}$; where H is the obtained result and U is the optimal.

Table 3.38. EDE QAP operational values

Parameter	Value
Strategy	1
CR	0.9
F	0.3

Table 3.39. EDE Irregular QAP comparison

Instant	flow dom	n	Optimal	TT	RTS	SA	GH	HAS-QAP	EDE
bur26a	2.75	26	5246670	0.208	-	0.1411	0.012	0	0.006
bur26b	2.75	26	3817852	0.441	-	0.1828	0.0219	0	0.0002
bur26c	2.29	26	5426795	0.17	-	0.0742	0	0	0.00005
bur26d	2.29	26	3821225	0.249	-	0.0056	0.002	0	0.0001
bur26e	2.55	26	5386879	0.076	-	0.1238	0	0	0.0002
bur26f	2.55	26	3782044	0.369	-	0.1579	0	0	0.000001
bur26g	2.84	26	10117172	0.078	-	0.1688	0	0	0.0001
bur26h	2.84	26	7098658	0.349	-	0.1268	0.0003	0	0.0001
chr25a	4.15	26	3796	15.969	16.844	12.497	2.6923	3.0822	0.227
els19	5.16	19	17212548	21.261	6.714	18.5385	0	0	0.0007
kra30a	1.46	30	88900	2.666	2.155	1.4657	0.1338	0.6299	0.0328
kra30b	1.46	30	91420	0.478	1.061	1.065	0.0536	0.0711	0.0253
tai20b	3.24	20	122455319	6.7	-	14.392	0	0.0905	0.0059
tai25b	3.03	25	344355646	11.486	-	8.831	0	0	0.003
tai30b	3.18	30	637117113	13.284	-	13.515	0.0003	0	0.0239
tai35b	3.05	35	283315445	10.165	-	6.935	0.1067	0.0256	0.0101
tai40b	3.13	40	637250948	9.612	-	5.43	0.2109	0	0.027
tai50b	3.1	50	458821517	7.602	-	4.351	0.2124	0.1916	0.001
tai60b	3.15	60	608215054	8.692	-	3.678	0.2905	0.0483	0.0144
tai80b	3.21	80	818415043	6.008	-	2.793	0.8286	0.667	0.0287

The comparison is done with Tabu Search (TT) [36], Reative Tabu Search (RTS) [1], Simulated Annealing (SA) [5], Genetic Hybrid (GH) [2] and Hybrid Ant Colony (HAS) [13].

Two trends are fairly obvious. The first is that for bur instances, HAS obtains the optimal, and is very closely followed by EDE by a margin of only 0.001 on average. For the *tai* instances, EDE competes very well, obtaining the best values for the larger problems and also obtains the best values for the *kra* problems. TT and RTS are shown to be not well adapted to irregular problems, producing 10% worse solution at times. GH which does not have memory retention capabilities does well, but does not produce optimal results with any regularity.

3.7.3 Experimentation for Regular QAP

The second section of QAP problems is discussed in this section. This is the set of regular problem as discussed in [13]. Regular problems are distinguished as having a flow−dominance of less than 1.2.

Comparison was done with the same heuristics as in the previous section. The results are presented in Table 3.40.

Three different set of instances are presented: *nug*, *sko*, *tai* and *wil*. Apart for the *nug20* instance, EDE finds the best solutions for all the reported instances. It can be observed that TT, GH and SA perform best for *sko* problems and RTS performs best for *tai* problems. On comparison with the optimal values, EDE obtains values with tolerance of only 0.01 on average for all instances.

A sample generation for *Bur26a* problem is given in Fig 3.18.

Table 3.40. EDE Regular QAP comparison

Instant	flow dom	n	Optimal	TT	RTS	SA	GH	HAS-QAP	EDE
nug20	0.99	20	2570	0	0.911	0.07	0	0	0.018
nug30	1.09	30	6124	0.032	0.872	0.121	0.007	0.098	0.005
sko42	1.06	42	15812	0.039	1.116	0.114	0.003	0.076	0.009
sko49	1.07	49	23386	0.062	0.978	0.133	0.04	0.141	0.009
sko56	1.09	56	34458	0.08	1.082	0.11	0.06	0.101	0.012
sko64	1.07	64	48498	0.064	0.861	0.095	0.092	0.129	0.013
sko72	1.06	72	66256	0.148	0.948	0.178	0.143	0.277	0.011
sko81	1.05	81	90998	0.098	0.88	0.206	0.136	0.144	0.011
tai20a	0.61	20	703482	0.211	0.246	0.716	0.628	0.675	0.037
tai25a	0.6	25	1167256	0.51	0.345	1.002	0.629	1.189	0.026
tai30a	0.59	30	1818146	0.34	0.286	0.907	0.439	1.311	0.018
tai35a	0.58	35	2422002	0.757	0.355	1.345	0.698	1.762	0.038
tai40a	0.6	40	3139370	1.006	0.623	1.307	0.884	1.989	0.032
tai50a	0.6	50	4941410	1.145	0.834	1.539	1.049	2.8	0.033
tai60a	0.6	60	7208572	1.27	0.831	1.395	1.159	3.07	0.037
tai80a	0.59	80	13557864	0.854	0.467	0.995	0.796	2.689	0.031
wil50	0.64	50	48816	0.041	0.504	0.061	0.032	0.061	0.004

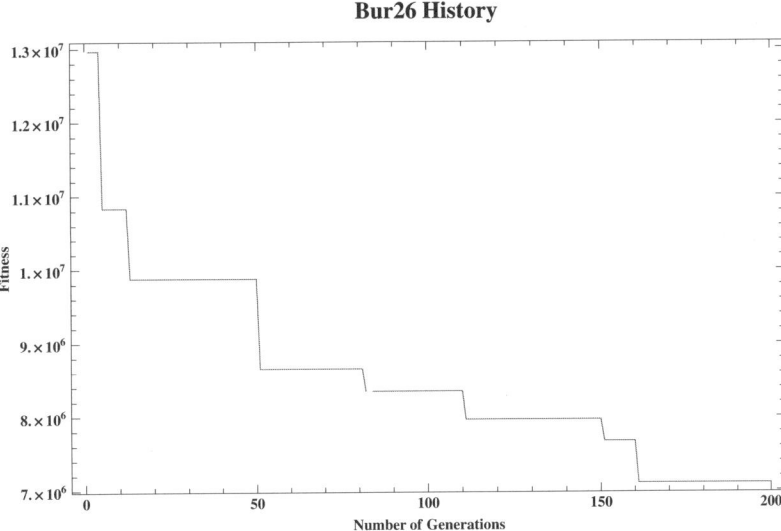

Fig. 3.18. Sample output of the Bur26a problem

3.8 Traveling Salesman Problem

The third and final problem class to be experimented is the *Traveling Salesman Problem* (TSP). The TSP is a very well known optimisation problem. A traveling salesman has a number, N, cities to visit. The sequence in which the salesperson visits different cities is called a *tour*. A tour is such that every city on the list is visited only once, except that the salesperson returns to the city from which it started. The objective to is minimise the total distance the salesperson travels, among all the tours that satisfy the criterion.

Several mathematical formulations exist for the TSP. One approach is to let x_{ij} be 1 if city j is visited immediately after i, and be 0 if otherwise [24, 25]. The formulation of TSP is given in Equations 3.17 to 3.20.

$$\min \sum_{i=1}^{N} \sum_{j=1}^{N} c_{ij} \bullet x_{ij} \qquad (3.17)$$

Each city is left after visiting subject to

$$\sum_{j=1}^{N} x_{ij} = 1; \forall i \qquad (3.18)$$

Ensures that each city is visited

$$\sum_{i=1}^{N} x_{ij} = 1; \forall j \qquad (3.19)$$

No subtours

$$x_{ij} = 0 \quad \text{or} \quad 1 \qquad (3.20)$$

Table 3.41. City distance matrix

City	A	B	C	D
E	2	3	2	4
D	1	5	1	
C	2	3		
B	1			

No subtours mean that there is no need to return to a city before visiting all the other cities. The objective function accumulates time as you go from city i to j. Constraint 3.18 ensures that the salesperson leaves each city. Constraint 3.19 ensures that the salesperson enters each city. A subtour occurs when the salesperson returns to a city prior to visiting all other cities. Restriction 3.20 enables the TSP formulation differs from the Linear Assignment Problem programming (LAP) formulation.

3.8.1 Traveling Salesman Problem Example

Assume there are five cities $\{A,B,C,D,E\}$, for a traveling salesman to visit as shown in Fig 3.19. The distance between each city is labelled in the vertex.

In order to understand TSP, assume a tour, where a salesman travels through all the cities and returns eventually to the original city. Such a tour can be given as $A \rightarrow B \rightarrow C \rightarrow D \rightarrow E \rightarrow A$. The graphical representation for such a tour is given in Fig 3.20.

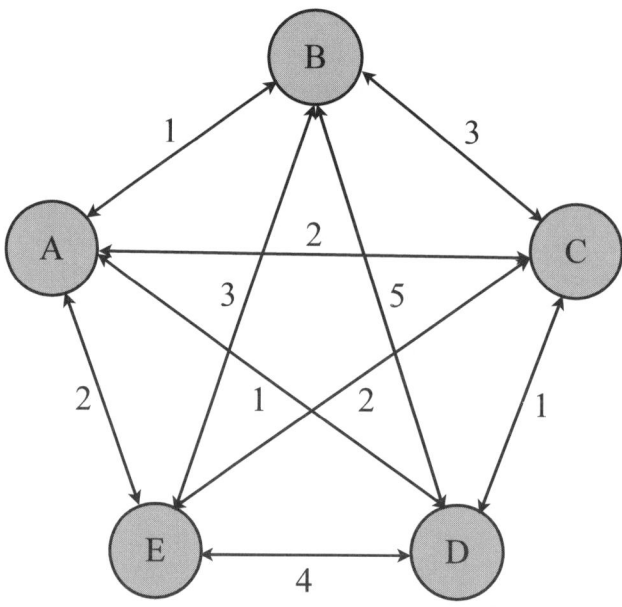

Fig. 3.19. TSP distance node graph

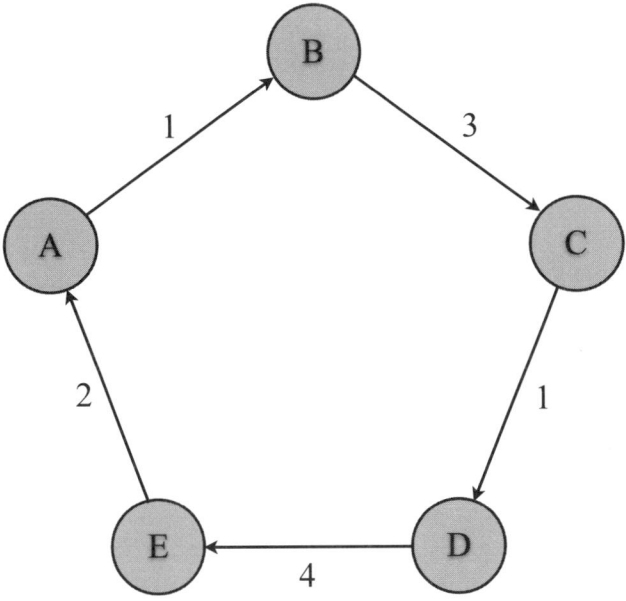

Fig. 3.20. Graphical representation for the tour $A \to B \to C \to D \to E \to A$

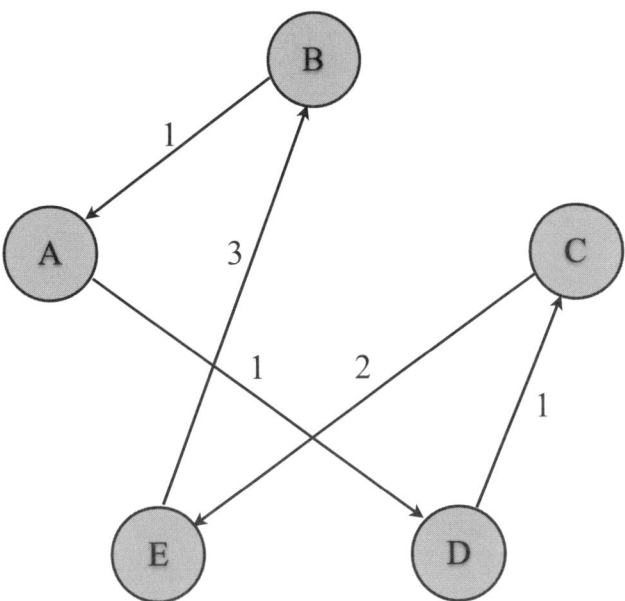

Fig. 3.21. Graphical representation for the tour $A \to D \to C \to E \to B \to A$

The total cost for this tour is 11.

The objective of TSP optimisation is to find a tour with the minimal value. Assume now another tour $A \to D \to C \to E \to B \to A$. The graphical representation is given in Fig 3.21.

The cost for this new tour is 8, which is a decrease from the previous tour of 11. This is now an improved tour. Likewise many other tours can be found which have better values.

3.8.2 Experimentation on Symmetric TSP

Symmetric TSP problem is one, where the distance between two cities is the same *to* and *fro*. This is considered the easiest branch of TSP problem.

The operational parameters for TSP is given in Table 3.42.

Experimentation was conducted on the *City* problem instances. These instances are of 50 cities and the results are presented in Table 3.43. Comparison was done with Ant Colony (ACS) [11], Simulated Annealing (SA) [21], Elastic Net (EN) [12], and Self Organising Map (SOM) [17]. The time values are presented alongside.

In comparison, ACS is the best performing heuristic for TSP. EDE performs well, with tolerance of 0.1 from the best performing heuristics on average.

3.8.3 Experimentation on Asymmetric TSP

Asymmetric TSP is the problem where the distance between the different cities is different, depending on the direction of travel. Five different instances were evaluated and compared with Ant Colony (ACS) with local search [11]. The experimetational results are given in Table 3.44.

Table 3.42. EDE TSP operational values

Parameter	Value
Strategy	9
CR	0.9
F	0.1

Table 3.43. EDE STSP comparison

Instant	ACS (average)	SA (average)	EN (average)	SOM (average)	EDE (average)
City set 1	5.88	5.88	5.98	6.06	5.98
City set 2	6.05	6.01	6.03	6.25	6.04
City set 3	5.58	5.65	5.7	5.83	5.69
City set 4	5.74	5.81	5.86	5.87	5.81
City set 5	6.18	6.33	6.49	6.7	6.48

Table 3.44. EDE ATSP comparison

Instant	Optimal	ACS 3-OPT best	ACS 3-OPT average	EDE
p43	5620	5620	5620	5639
ry48p	14422	14422	14422	15074
ft70	38673	38673	38679.8	40285
kro124p	36230	36230	36230	41180
ftv170	2755	2755	2755	6902

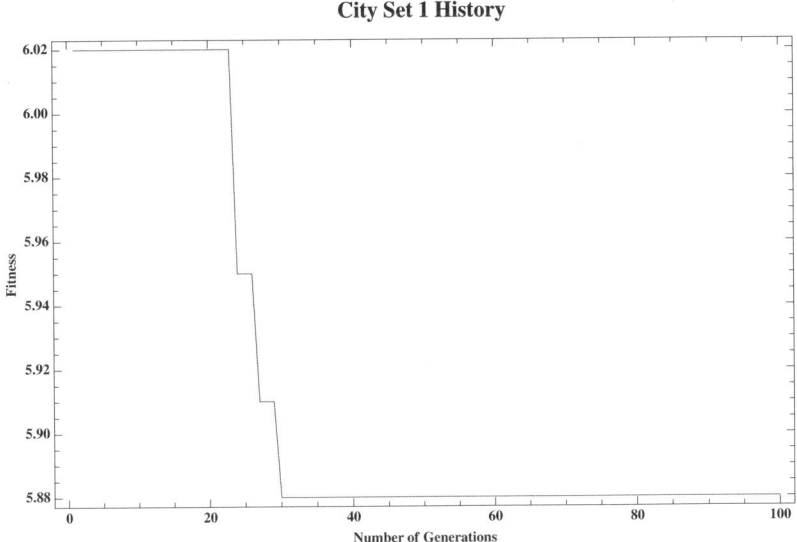

Fig. 3.22. Sample output of the City set 1 problem

ACS heuristic performs very well, obtaining the optimal value, whereas EDE has an average performance. The difference is that ACS employs 3−Opt local search on each generation of its best solution, where as EDE has a 2−Opt routine valid only in local optima stagnation.

A sample generation for *City set 1* problem is given in Fig 3.22.

3.9 Analysis and Conclusion

One the few ways in which the validation of a permutative approach for a real domain based heuristic can be done is empirically; through expensive experimentation's across different problem classes, as attempted here. Through the results obtained, it can be stated that EDE is a valid approach for permutative problems. One of the differing evident features, is that the operating parameters for each class of problems is unique. No definite conclusions can be made on this aspect, apart from the advise for simulations for tuning.

Another important feature of EDE is the level of stochasticity. DE has two levels; first the initial population and secondly the crossover. EDE has five; in addition to the two mentioned, the third is repairment, the fourth is mutation and fifth is crossover. All these three are embedded on top of the DE routine, so the DE routines are a directive search guide with refinement completed in the subsequent routines.

Local search was included in EDE because permutative problems usually require triangle inequality routines. TSP is notorious in this respect, and most heuristics have to employ local search in order to find good solutions. ACS [11], Scatter Search [14] apply local search on each and every solution. This increases computational time and reduces effectiveness of the heuristic for practical applications. The idea of EDE was to only employ local search when stagnation is detected, and to employ the simplest and time economical one.

In terms of produced results, EDE is effective, and more so since it was left in non-altered form for all the problem classes. This is a very important feature since it negates re-programming for other problem instances. Another important feature is that EDE is fairly fast for these problems. Naturally, the increase in problem size increases the execution time, however EDE does not employ any analytical formulation within its heuristic, which keeps down the execution time while producing the same results as with other heuristics.

It is hoped that the basic framework of this approach will be improved to include more problem instances, like Job Shop Scheduling and other manufacturing scheduling problems.

References

1. Battitti, R., Tecchiolli, G.: The reactive tabu search. ORCA Journal on Computing 6, 126–140 (1994)
2. Burkard, R., Rendl, F.: A thermodynamically motivated simulation procedure for combinatorial optimisation problems. Eur. J. Oper. Res. 17, 169–174 (1994)
3. Carlier, J.: Ordonnancements a Contraintes Disjonctives. RAIRO. Oper. Res. 12, 333–351 (1978)
4. Chen, C., Vempati, V., Aljaber, N.: An application of genetic algorithms for the flow shop problems. Eur. J. Oper. Res. 80, 359–396 (1995)
5. Connolly, D.: An improved annealing scheme for the QAP. Eur. J. Oper. Res. 46, 93–100 (1990)
6. Davendra, D.: Differential Evolution Algorithm for Flow Shop Scheduling, Bachelor Degree Thesis, University of the South Pacific (2001)
7. Davendra, D.: Hybrid Differential Evolution Algorithm for Discrete Domain Problems. Master Degree Thesis, University of the South Pacific (2003)
8. Davendra, D., Onwubolu, G.: Flow Shop Scheduling using Enhanced Differential Evolution. In: Proceeding of the 21st European Conference on Modelling and Simulation, Prague, Czech Republic, June 4-5, pp. 259–264 (2007)
9. Davendra, D., Onwubolu, G.: Enhanced Differential Evolution hybrid Scatter Search for Discrete Optimisation. In: Proceeding of the IEEE Congress on Evolutionary Computation, Singapore, September 25-28, pp. 1156–1162 (2007)

10. Dorigo, M., Maniezzo, V., Colorni, A.: The Ant System: optimisation by a colony of co-operating agents. IEEE Trans. Syst. Man Cybern B Cybern. 26(1), 29–41 (1996)
11. Dorigo, M., Gambardella, L.: Ant Colony System: A Co-operative Learning Approach to the Traveling Salesman Problem. IEEE Trans. Evol. Comput. 1, 53–65 (1997)
12. Durbin, R., Willshaw, D.: An analogue approach to the travelling salesman problem using the elastic net method. Nature 326, 689–691 (1987)
13. Gambardella, L., Thaillard, E., Dorigo, M.: Ant Colonies for the Quadratic Assignment Problem. Int. J. Oper. Res. 50, 167–176 (1999)
14. Glover, F.: A template for scatter search and path relinking. In: Hao, J.-K., Lutton, E., Ronald, E., Schoenauer, M., Snyers, D. (eds.) AE 1997. LNCS, vol. 1363, pp. 13–54. Springer, Heidelberg (1998)
15. Heller, J.: Some Numerical Experiments for an MJ Flow Shop and its Decision- Theoretical aspects. Oper. Res. 8, 178–184 (1960)
16. Ho, Y., Chang, Y.-L.: A new heuristic method for the n job, m - machine flow-shop problem. Eur. J. Oper. Res. 52, 194–202 (1991)
17. Kara, L., Atkar, P., Conner, D.: Traveling Salesperson Problem (TSP) Using Shochastic Search. In: Advanced AI Assignment, Carnegie Mellon Assignment, Pittsbergh, Pennsylvania, p. 15213 (2003)
18. Koopmans, T., Beckmann, M.: Assignment problems and the location of economic activities. Econometrica 25, 53–76 (1957)
19. Lampinen, J., Zelinka, I.: Mechanical engineering design optimisation by Differential evolution. In: Corne, D., Dorigo, M., Glover, F. (eds.) New Ideas in Optimisation, pp. 127–146. McGraw Hill, International, UK (1999)
20. Lawler, E., Lensta, J., Rinnooy, K., Shmoys, D.: Sequencing and scheduling: algorithms and complexity. In: Graves, S., Rinnooy, K., Zipkin, P. (eds.) Logistics of Production and Inventory, pp. 445–522. North Holland, Amsterdam (1995)
21. Lin, F., Kao, C., Hsu: Applying the genetic approach to simulated annealing in solving NP-hard problems. IEEE Trans. Syst. Man Cybern. B Cybern. 23, 1752–1767 (1993)
22. Nowicki, E., Smutnicki, C.: A fast tabu search algorithm for the permutative flow shop problem. Eur. J. Oper. Res. 91, 160–175 (1996)
23. Onwubolu, G.: Optimisation using Differential Evolution Algorithm. Technical Report TR-2001-05, IAS (October 2001)
24. Onwubolu, G.: Emerging Optimisation Techniques in Production Planning and Control. Imperial Collage Press, London (2002)
25. Onwubolu, G., Clerc, M.: Optimal path for automated drilling operations by a new heuristic approach using particle swarm optimisation. Int. J. Prod. Res. 42(3), 473–491 (2004)
26. Onwubolu, G., Davendra, D.: Scheduling flow shops using differential evolution algorithm. Eur. J. Oper. Res. 171, 674–679 (2006)
27. Onwubolu, G., Kumalo, T.: Optimisation of multi pass tuning operations with genetic algorithms. Int. J. Prod. Res. 39(16), 3727–3745 (2001)
28. Onwubolu, G., Mutingi, M.: Genetic algorithm for minimising tardiness in flow-shop scheduling. Prod. Plann. Contr. 10(5), 462–471 (1999)
29. Operations Reserach Library (Cited September 13, 2008),
    http://people.brunel.ac.uk/~mastjjb/jeb/info.htm
30. Pinedo, M.: Scheduling: theory, algorithms and systems. Prentice Hall, Inc., New Jersey (1995)
31. Price, K.: An introduction to differential evolution. In: Corne, D., Dorigo, M., Glover, F. (eds.) New Ideas in Optimisation, pp. 79–108. McGraw Hill, International (1999)
32. Price, K., Storn, R.: Differential evolution homepage (2001) (Cited September 10, 2008),
    http://www.ICSI.Berkeley.edu/~storn/code.html

33. Reeves, C.: A Genetic Algorithm for Flowshop Sequencing. Comput. Oper. Res. 22, 5–13 (1995)
34. Reeves, C., Yamada, T.: Genetic Algorithms, path relinking and flowshop sequencing problem. Evol. Comput. 6, 45–60 (1998)
35. Sahni, S., Gonzalez, T.: P-complete approximation problems. J. ACM 23, 555–565 (1976)
36. Taillard, E.: Robust taboo search for the quadratic assignment problem. Parallel Comput. 17, 443–455 (1991)
37. Taillard, E.: Benchmarks for basic scheduling problems. Eur. J. Oper. Res. 64, 278–285 (1993)
38. Tasgetiren, M., Liang, Y.-C., Sevkli, M., Gencyilmaz, G.: Differential Evolution Algorithm for Permutative Flowshops Sequencing Problem with Makespan Criterion. In: 4th International Symposium on Intelligent Manufacturing Systems, IMS 2004, Sakaraya, Turkey, September 5–8, pp. 442–452 (2004)
39. Widmer, M., Hertz, A.: A new heuristic method for the flow shop sequencing problem. Eur. J. Oper. Res. 41, 186–193 (1989)

4

Relative Position Indexing Approach

Daniel Lichtblau

Wolfram Research, Inc., 100 Trade Center Dr., Champaign, IL 61820, USA
danl@wolfram.com

Abstract. We introduce some standard types of combinatorial optimization problems, and indicate ways in which one might attack them using Differential Evolution. Our main focus will be on *indexing by relative position* (also known as *order based representation*); we will describe some related approaches as well. The types of problems we will consider, which are abstractions of ones from engineering, go by names such as *knapsack problems*, *set coverings*, *set partitioning*, and *permutation assignment*. These are historically significant types of problems, as they show up frequently, in various guises, in engineering and elsewhere. We will see that a modest amount of programming, coupled with a sound implementation of Differential Evolution optimization, can lead to good results within reasonable computation time. We will also show how Differential Evolution might be hybridized with other methods from combinatorial optimization, in order to obtain better results than might be found with the individual methods alone.

4.1 Introduction

The primary purpose of this chapter is to introduce a few standard types of combinatorial optimization problems, and indicate ways in which one might attack them using Differential Evolution. Our main focus will be on *indexing by relative position* (also known as *order based representation*); we will describe some related approaches as well. We will not delve much into why these might be regarded as "interesting" problems, as that would be a chapter– or, more likely, book– in itself. Suffice it to say that many problems one encounters in the combinatorial optimization literature have their origins in very real engineering problems, e.g. layout of hospital wings, electronic chip design, optimal task assignments, boolean logic optimization, routing, assembly line design, and so on. The types of problems we will consider, which are abstractions of the ones from engineering, go by names such as *knapsack problems*, *set coverings*, *set partitioning*, and *permutation assignment*. A secondary goal of this chapter will be to introduce a few ideas regarding hybridization of Differential Evolution with some other methods from optimization.

I will observe that, throughout this chapter at least, we regard Differential Evolution as a *soft* optimization tool. Methods we present are entirely heuristic in nature. We usually do not get guarantees of result quality; generally this must be assessed by independent means (say, comparison with other tactics such as random search or greedy algorithms, or a priori problem-specific knowledge). So we use the word *optimization* a bit loosely, and really what we usually mean is *improvement*. While this may seem to be bad from a theoretical point of view, it has advantages. For one, the field is relatively young and is amenable to an engineering mind set. One need not invent a new branch of mathematics in order to make progress. At the end of the chapter we will see a few directions that might, for the ambitious reader, be worthy of further pursuit.

Another caveat is that I make no claim as to Differential Evolution being the *best* method for the problems we will discuss. Nor do I claim that the approaches to be seen are the only, let alone best, ways to use it on these problems (if that were the case, this would be a very short book indeed). What I do claim is that Differential Evolution is quite a versatile tool, one that can be adapted to get reasonable results on a wide range of combinatorial optimization problems. Even more, this can be done using but a small amount of code. It is my hope to convey the utility of some of the methods I have used with success, and to give ideas of ways in which they might be further enhanced.

As I will illustrate the setup and solving attempts using *Mathematica* [13], I need to describe in brief how Differential Evolution is built into and accessed within that program. Now recall the general setup for this method. We have some number of vectors, or *chromosomes*, of continuous-valued *genes*. They *mate* according to a crossover probability, *mutate* by differences of distinct other pairs in the pool, and compete with a parent chromosome to see who moves to the next generation. All these are as described by Price and Storn, in their Dr. Dobbs Journal article from 1997 [10]. In particular, crossover and mutation parameters are as described therein. In *Mathematica* the relevant options go by the names of CrossProbability, ScalingFactor, and SearchPoints. Each variable corresponds to a gene on every chromosome. Using the terminology of the article, CrossProbability is the CR parameter, SearchPoints corresponds to NP (size of the population, that is, number of chromosome vectors), and ScalingFactor is F. Default values for these parameters are roughly as recommended in that article.

The function that invokes these is called NMinimize. It takes a Method option that can be set to DifferentialEvolution. It also takes a MaxIterations option that, for this method, corresponds to the number of generations. Do not be concerned if this terminology seems confusing. Examples to be shown presently will make it all clear.

One explicitly invokes Differential Evolution in *Mathematica* as follows.

NMinimize[objective, constraints,
    variables, Method → {"DifferentialEvolution", methodoptions}, otheropts]

Here methodoptions might include setting to nondefault values any or all of the options indicated below. We will show usage of some of them as we present examples. Further details about these options may be found in the program documentation. All of which is available online; see, for example,
http://reference.wolfram.com/mathematica/ref/NMinimize.html

Here are the options one can use to control behavior of NMinimize. Note that throughout this chapter, code input is in **bold face**, and output, just below the input, is not.

Options[NMinimizeDifferentialEvolution]

{*CrossProbability* → $\frac{1}{2}$, InitialPoints → Automatic,
    PenaltyFunction → Automatic, PostProcess → Automatic,

RandomSeed $\to$ 0, ScalingFactor $\to \frac{3}{5}$,
SearchPoints $\to$ Automatic, Tolerance $\to$ 0.001}

There are some issues in internal implementation that need discussion to avoid later confusion. First is the question of how constraints are enforced. This is particularly important since we will often constrain variables to take on only integer values, and in specific ranges. For integrality enforcement there are (at least) two viable approaches. One is to allow variables to take on values in a continuum, but use penalty functions to push them toward integrality [2]. For example, one could add, for each variable x, a penalty term of the form $(x - \text{round}(x))^2$ (perhaps multiplied by some suitably large constant). NMinimize does not use this approach, but a user can choose to assist it to do so by explicitly using the PenaltyFunction method option. Another method, the one used by NMinimize, is to explicitly round all (real-valued) variables before evaluating in the objective function. Experience in the development of this function indicated this was typically the more successful approach.

This still does not address the topic of range enforcement. For example, say we are using variables in the range $\{1,\ldots,n\}$ to construct a permutation of n elements. If a value slips outside the range then we might have serious difficulties. For example, in low level programming languages such as C, having an out-of-bounds array reference can cause a program to crash. While this would not as likely happen in *Mathematica*, the effect would still be bad, for example a hang or garbage result due to processing of a meaningless symbolic expression. So it is important either that our code, or NMinimize, carefully enforce variable bounds. As it happens, the implementation does just this. If a variable is restricted to lie between a low and high bound (this is referred to as a *rectangular*, or *box*, constraint), then the NMinimize code will force it back inside the boundary. Here I should mention that this is really a detail of the implementation, and should not in general be relied upon by the user. I point it out so that the reader will not be mystified upon seeing code that blithely ignores the issue throughout the rest of this chapter. I also note that it is not hard to make alterations e.g. to an objective function, to preprocess input so that bound constraints are observed; in order to maximize simplicity of code, I did not do this.

I will make a final remark regarding use of *Mathematica* before proceeding to the material of this chapter. It is not expected that readers are already familiar with this program. A consequence is that some readers will find parts of the code we use to be less than obvious. This is to be expected any time one first encounters a new, complicated computer language. I will try to explain in words what the code is doing. Code will also be preceded by a concise description, in outline form, that retains the same order as the code itself and thus serves as a form of pseudocode. The code details are less important. Remember that the emphasis is on problem solving approaches using Differential Evolution in general; specifics of a particular language or implementation, while of independent interest, take a back seat to the Big Picture.

In the literature on combinatorial (and other) optimization via evolutionary means, one frequently runs across notions of *genotype* and *phenotype*. The former refers to the actual chromosome values. Recall the basic workings of Differential Evolution. One typically forms a new chromosome from mating its parent chromosome with a mutation of a random second parent. Said mutation is in turn given as a difference of two other random chromosomes. These operations are all done at the genotype level. It is in translating the chromosome to a combinatorial object e.g. a permutation, that one encounters the phenotype. This refers, roughly, to the *expression* of the chromosome as something we can use for objective function evaluation. Said slightly differently, one *decodes* a genotype to obtain a phenotype. We want genotypes that are amenable to the mutation and mating operations of Differential Evolution, and phenotypes that will respond well to the genotype, in the sense of allowing for reasonable improvement of objective function. Discussion of these matters, with respect to the particulars of Differential Evolution, may be found in [11]. Early discussion of these issues, and methods for handling them, appear in [4] and [3].

4.2 Two Simple Examples

I like to start discussion of Differential Evolution in discrete optimization by presenting two fairly straightforward examples. They serve to get the reader acclimated to how we might set up simple problems, and also to how they look as input to *Mathematica*. These are relatively simple examples of discrete optimization, not involving combinatorial problems, and hence are good for easing into the main material of this chapter.

4.2.1 Pythagorean Triples

First we will search for Pythagorian triples. These, as one may recall from high school, are integer triples (x, y, z) such that $x^2 + y^2 = z^2$. So we wish to find integer triples that satisfy this equation. One way to set up such a problem is to form the square of the difference, $x^2 + y^2 - z^2$. We seek integer triples that make this vanish, and moreover this vanishing is a minimization condition (because we have a square). Note that this is to some extent arbitrary, and minimizing the absolute value rather than the square would suffice just as well for our purpose.

We constrain all variables to be between 5 and 25 inclusive. We also specify explicitly that the variables are integer valued. We will say a bit more about this in a moment.

NMinimize[{$(x^2 + y^2 - z^2)^2$, Element[{x, y, z}, Integers],
  $5 \le x \le 25, 5 \le y \le 25, 5 \le z \le 25, x \le y$}, {$x, y, z$}]

{0., {$x \to 7, y \to 24, z \to 25$}}

We see that `NMinimize` is able to pick an appropriate method by default. Indeed, it uses `DifferentialEvolution` when variables are specified as discrete, that is, integer valued.

Now we show how to obtain different solutions by specifying that the random seed used by the `DifferentialEvolution` method change for each run. We will suppress warning messages (the algorithm mistakenly believes it is not converging). After all, we are only interested in the results; we can decide for ourselves quite easily if they work.

Quiet[
 Timing[
  Table[NMinimize[$\{(x^2+y^2-z^2)^2$,
   Element[$\{x,y,z\}$, Integers], $5 \le x \le 25, 5 \le y \le 25, 5 \le z \le 25, x \le y\}$,
   $\{x,y,z\}$,
   Method $\to$ "DifferentialEvolution", RandomSeed $\to$ RandomInteger[1000]],
   $\{20\}$]]]

$\{17.1771, \{\{0., \{x \to 9, y \to 12, z \to 15\}\}, \{0., \{x \to 15, y \to 20, z \to 25\}\},$
$\{0., \{x \to 6, y \to 8, z \to 10\}\}, \{0., \{x \to 5, y \to 12, z \to 13\}\},$
$\{0., \{x \to 6, y \to 8, z \to 10\}\}, \{0., \{x \to 7, y \to 24, z \to 25\}\},$
$\{0., \{x \to 15, y \to 20, z \to 25\}\}, \{0., \{x \to 15, y \to 20, z \to 25\}\},$
$\{0., \{x \to 15, y \to 20, z \to 25\}\}, \{0., \{x \to 8, y \to 15, z \to 17\}\},$
$\{0., \{x \to 5, y \to 12, z \to 13\}\}, \{0., \{x \to 9, y \to 12, z \to 15\}\},$
$\{0., \{x \to 9, y \to 12, z \to 15\}\}, \{0., \{x \to 5, y \to 12, z \to 13\}\},$
$\{0., \{x \to 6, y \to 8, z \to 10\}\}, \{0., \{x \to 5, y \to 12, z \to 13\}\},$
$\{0., \{x \to 5, y \to 12, z \to 13\}\}, \{0., \{x \to 5, y \to 12, z \to 13\}\},$
$\{0., \{x \to 5, y \to 12, z \to 13\}\}, \{0., \{x \to 15, y \to 20, z \to 25\}\}\}\}$

We observe that each of these is a valid Pythagorean triple (of course, there are several repeats). Recalling our objective function, any failure would appear as a false minimum, that is to say, a square integer strictly larger than zero.

4.2.1.1 A Coin Problem

We start with a basic coin problem. We are given 143,267 coins in pennies, nickels, dimes, and quarters, of total value $12563.29, and we are to determine how many coins might be of each type. There are several ways one might set up such a problem in NMinimize. We will try to minimize the sum of squares of differences between actual values and desired values of the two linear expressions implied by the information above. For our search space we will impose obvious range constraints on the various coin types. In order to obtain different results we will want to alter the seeding of the random number generator; this changes the random initial parameters used to seed the optimization code. That is why we specify the method with this option added. We will do 10 runs of this.

Timing[Table[
 $\{$min, sol$\}$ = NMinimize[
  $\{(p+5n+10d+25q-1256329)^2 + (p+n+d+q-143267)^2$,
   $\{p,n,d,q\} \in$ Integers, $0 \le p \le 1256329, 0 \le n \le 1256329/5$,
   $0 \le d \le 1256329/10, 0 \le q \le 1256329/25\}$,
   $\{p,n,d,q\}$, MaxIterations $\to$ 1000,
   Method $\to$ $\{$DifferentialEvolution, RandomSeed $\to$ Random[Integer, 1000]$\}$],
   $\{10\}$]]

NMinimize::cvmit : Failed to converge to the
requested accuracy or precision within 1000 iterations.

$\{229.634, \{\{0., \{p \to 22554, n \to 70469, d \to 24978, q \to 25266\}\},$
$\{0., \{p \to 4094, n \to 79778, d \to 42102, q \to 17293\}\},$
$\{0., \{p \to 23139, n \to 64874, d \to 31502, q \to 23752\}\},$
$\{0., \{p \to 26649, n \to 72620, d \to 15558, q \to 28440\}\},$
$\{0., \{p \to 2914, n \to 76502, d \to 48358, q \to 15493\}\},$
$\{0., \{p \to 9714, n \to 49778, d \to 73110, q \to 10665\}\},$
$\{0., \{p \to 26019, n \to 26708, d \to 77782, q \to 12758\}\},$
$\{0., \{p \to 58229, n \to 31772, d \to 19494, q \to 33772\}\},$
$\{0., \{p \to 8609, n \to 70931, d \to 46674, q \to 17053\}\},$
$\{0., \{p \to 35049, n \to 55160, d \to 25398, q \to 27660\}\}\}\}$

We obtained valid solutions each time. Using only, say, 400 iterations we tend to get solutions about half the time and "near" solutions the other half (wherein either the number of coins and/or total value is off by a very small amount). Notice that this type of problem is one of constraint satisfaction. An advantage to such problems is that we can discern from the proposed solution whether it is valid; those are exactly the cases for which we get an object value of zero, with all constraints satisfied.

4.3 Maximal Determinants

In this section we illustrate a heuristic methods on certain extremal matrix problems of modest size. As motivation for looking at this particular problem, I remark that it is sometimes important to understand extremal behavior of random polynomials or matrices comprised of elements from a given set.

Below we apply knapsack-style optimization to study determinants of matrices of integers with all elements lying in the set $\{-1,0,1\}$. The problem is to minimize the determinant of such a matrix (since we can multiply any row by -1 and still satisfy the constraints, the smallest negative value corresponds to the largest positive value). We will make the simplifying assumption that all diagonal elements are 1. Strictly speaking this is not combinatorial optimization, but it is a close relative, and will help to get the reader acquainted with the programming commands we will be using in this chapter. Thus is also a good example with which to begin this chapter.

Our objective function is simply the determinant. We want it only to evaluate when the variables have been assigned numeric values. This is quite important because symbolic determinants are quite slow to compute. So we set up the function so that it is only defined when numeric values are plugged in.

detfunc[a : {{_?NumberQ..}..}]/;Length[a] == Length[First[a]]:=Det[a]

Our code will take a matrix dimension as argument, and also an optional argument specifying whether to print the constraints. We use that in a small problem to show the constraints explicitly, so that the reader may check that we have set this up correctly. Before showing the actual code we first outline the process.

---

*Outline of **detMin***

1. Input: the dimension, and the parameter settings we will use for NMinimize.
2. Create a matrix of variables.
3. Create a set of constraints.
    All variables must be integers.
    All variables lie in the range $[-1, 1]$.
    Variables corresponding to the diagonal elements are all set to 1.
4. Call NMinimize on the objective function, using the above constraints and taking program parameters from the argument list.
5. Return the optimum found by NMinimize, along with the matrix that gives this value.

Here is the actual program to do this.

detMin[n_, cp_, sp_, it_, printsetup_:False]:=Module[
{mat, vars, problemlist, j, best},
mat = Array[x, {n,n}];
vars = Flatten[mat];
problemlist =
  {detfunc[mat], Flatten[{Element[vars, Integers], Map[$-1 \leq \# \leq 1$&, vars],
    Table[$x[j,j] == 1$, {j,n}]}]};
If[printsetup, Print[problemlist[[2]]]];
best = NMinimize[problemlist, vars, MaxIterations $\rightarrow$ it,
Method $\rightarrow$ {DifferentialEvolution, CrossProbability $\rightarrow$ cp, SearchPoints $\rightarrow$ sp}];
{best[[1]], mat/.best[[2]]}
]

Here is our result for three-by-three matrices. We also show the constraints for this small example.

Timing[{min, mat} = detMin[3, .1, 20, 20, True]]

$\{(x[1,1] | x[1,2] | x[1,3] | x[2,1] | x[2,2] | x[2,3] | x[3,1] | x[3,2] | x[3,3]) \in$ Integers,
  $-1 \leq x[1,1] \leq 1, -1 \leq x[1,2] \leq 1, -1 \leq x[1,3] \leq 1,$
  $-1 \leq x[2,1] \leq 1, -1 \leq x[2,2] \leq 1, -1 \leq x[2,3] \leq 1,$
  $-1 \leq x[3,1] \leq 1, -1 \leq x[3,2] \leq 1, -1 \leq x[3,3] \leq 1,$
  $x[1,1] == 1, x[2,2] == 1, x[3,3] == 1\}$

$\{0.528033, \{-4., \{\{1,1,1\}, \{1,1,-1\}, \{1,-1,1\}\}\}\}$

We obtain -4 as the minimum (can you do better?) We now try at dimension 7. We will use a larger search space and more iterations. Indeed, our option settings were determined by trial and error. Later we will say more about how this might systematically be done.

Timing[{min, mat} = detMin[7, .1, 80, 80]]

$\{54.6874, \{-576., \{\{1, 1, -1, -1, 1, -1, 1\}, \{1, 1, -1, 1, -1, -1, -1\},$
$\{1, -1, 1, 1, 1, -1, -1\}, \{-1, -1, -1, 1, 1, 1, 1\}, \{1, 1, -1, -1, 1, 1, -1\},$
$\{1, 1, 1, 1, 1, 1, 1\}, \{1, -1, -1, -1, -1, 1, 1\}\}\}\}$

Readers familiar with the *Hadamard bound* for absolute values of matrix determinants will recognize that the minimum must be no smaller than the ceiling of $-7^{\frac{7}{2}}$, or -907. (In brief, this bound is the product of the lengths of the rows of a matrix; for our family, the maximal length of each row is $\sqrt{7}$. That this product maximizes the absolute value of the determinant can be observed from the fact that this absolute value is the volume of the rectangular prism formed by the row vectors of the matrix. This volume can be no larger than the product of their lengths; it achieves that value precisely when the rows are pairwise orthogonal.)

We can ask how good is the quality of our result. Here is one basis for comparison. A random search that took approximately twice as long as the code above found nothing smaller than -288. Offhand I do not know if -576 is the true minimum, though I suspect that it is.

It is interesting to see what happens when we try this with dimension increased to eight.

Timing[{min, mat} = detMin[8, 1/50, 100, 200]]

$\{222.618, \{-4096., \{\{1, -1, 1, 1, 1, -1, -1, -1\}, \{-1, 1, 1, -1, 1, 1, 1, -1, -1\},$
$\{-1, 1, 1, 1, 1, -1, 1, 1\}, \{1, 1, -1, 1, 1, -1, -1, -1, 1\},$
$\{-1, -1, -1, -1, 1, -1, -1, 1\}, \{1, 1, 1, -1, 1, 1, -1, 1\},$
$\{1, 1, -1, -1, 1, -1, 1, -1\}, \{1, -1, -1, 1, 1, 1, 1, 1\}\}\}\}$

In this case we actually attained the Hadamard bound; one can check that the rows (and likewise the columns) are all pairwise orthogonal, as must be the case in order to attain the Hadamard bound. Indeed, when the dimension is a power of two, one can always attain this bound. The motivated reader might try to work out a recursive (or otherwise) construction that gives such pairwise orthogonal sets.

4.4 Partitioning a Set

The last sections were a warmup to the main focus of this chapter. We introduced a bit of *Mathematica* coding, and in particular use of Differential Evolution, in the contect of discrete optimization. We now get serious in discussing combinatorial optimization problems and techniques.

4 Relative Position Indexing Approach

We start with the *Set Partitioning Problem*. We will illustrate this with an old example from computational folklore: we are to partition the integers from 1 to 100 into two sets of 50, such that the sums of the square roots in each set are as close to equal as possible.

There are various ways to set this up as a problem for NMinimize. We will show two of them. First we will utilize a simple way of choosing 50 elements from a set of 100. We will use 100 real values, all between 0 and 1. (Note that we are using continuous variables even though the problem itself involves a discrete set.) We take their *relative positions* as defining a permutation of the integers from 1 to 100. A variant of this approach to decoding permutations is described in [4, 3].

In more detail: their sorted ordering (obtained, in our code, from the *Mathematica* Ordering function) determines which is to be regarded as first, which as second, and so on. As this might be confusing, we illustrate the idea on a smaller set of six values. We begin with our range of integers from 1 to 6.

smallset = Range[6]

$\{1,2,3,4,5,6\}$

Now suppose we also have a set of six real values between 0 and 1.

vals = RandomReal[1, {6}]

$\{0.131973, 0.80331, 0.28323, 0.694475, 0.677346, 0.255748\}$

We use this second set of values to split smallset into two subsets of three, simply by taking as one such subset the elements with positions corresponding to those of the three smallest member of vals. The complementary subset would therefore be the elements with positions corresponding to those of the three largest members of vals. One can readily see (and code below will confirm) that the three smallest elements of vals, in order of increasing size, are the first, sixth, and third elements.

Ordering[vals]

$\{1,6,3,5,4,2\}$

We split this into the positions of the three smallest, and those of the three largest, as below.

{smallindices, largeindices} = {Take[#, 3], Drop[#, 3]}&[Ordering[vals]]

$\{\{1,6,3\},\{5,4,2\}\}$

We now split smallset according to these two sets of indices. Because it is simply the values one through six, the subsets are identical to their positions.

{s1, s2} = Map[smallset[[#]]&, {smallindices, largeindices}]

$\{\{1,6,3\},\{5,4,2\}\}$

The same idea applies to splitting any set of an even number of elements (small modifications could handle an odd number, or a split into subsets of unequal lengths).

With this at hand we are now ready to try our first method for attacking this problem.

4.4.1 Set Partitioning via Relative Position Indexing

Here is the code we actually use to split our 100 integers into two sets of indices.

Outline of splitRange

1. Input: a vector of real numbers, of even length.
2. Return the positions of the smaller half of elements, followed by those of the larger half.

---

splitRange[vec_]:=With[
 {newvec = Ordering[vec], halflen = Floor[Length[vec]/2]},
 {Take[newvec, halflen], Drop[newvec, halflen]}]

Just to see that it works as advertised, we use it to replicate the result from our small example above.

splitRange[vals]

$\{\{1, 6, 3\}, \{5, 4, 2\}\}$

Once we have a way to associate a pair of subsets to a given set of 100 values in the range from 0 to 1, we form our objective function. A convenient choice is simply an absolute value of a difference; this is often the case in optimization problems. We remark that squares of differences are also commonly used, particularly when the optimization method requires differentiability with respect to all program variables. This is not an issue for Differential Evolution, as it is a derivative-free optimization algorithm.

Here is an outline of the objective function, followed by the actual code.

Outline of spfun

1. Input: a vector of real numbers, of even length.
2. Use splitRange to find positions of the smaller half of elements, and the positions of the larger half.
3. Sum the square roots of the first set of positions, and likewise sum the square roots of the second set.
4. Return the absolute value of the difference of those two sums.

---

spfun[vec : {_Real}]:=
 With[{vals = splitRange[vec]},
  Abs[(Apply[Plus, Sqrt[N[First[vals]]]] − Apply[Plus, Sqrt[N[Last[vals]]]])]]

It may be a bit difficult to see what this does, so we illustrate again on our small example. Supposing we have split smallset into two subsets as above, what is the objective function? Well, what we do is take the first, sixth, and third elements, add their square roots, and do likewise with the fifth, fourth, and second elements. We subtract one of these sums from the other and take the absolute value of this difference. For speed we do all of this in machine precision arithmetic. In exact form it would be:

sqrts = Sqrt[splitRange[vals]]

$$\left\{\left\{1, \sqrt{6}, \sqrt{3}\right\}, \left\{\sqrt{5}, 2, \sqrt{2}\right\}\right\}$$

sums = Total[sqrts, {2}]

$$\left\{1 + \sqrt{3} + \sqrt{6}, 2 + \sqrt{2} + \sqrt{5}\right\}$$

sumdifference = Apply[Subtract, sums]

$$-1 - \sqrt{2} + \sqrt{3} - \sqrt{5} + \sqrt{6}$$

abssummdiffs = Abs[sumdifference]

$$1 + \sqrt{2} - \sqrt{3} + \sqrt{5} - \sqrt{6}$$

approxabs = N[abssummdiffs]

0.468741

As a check of consistency, observe that this is just what we get from evaluating our objective function on vals.

spfun[vals]

0.468741

We now put these components together into a function that provides our set partition.

---

*Outline of **getHalfSet***

1. Input: An even integer n, and options to pass along to NMinimize.
2. Create a list of variables, vars, of length n.
3. Set up initial ranges that the variables all lie between 0 and 1 (these are not hard constraints but just tell NMinimize where to take random initial values).

> 4. Call `NMinimize`, passing it `obfun[vars]` as objective function.
> 5. Return the minimum value found, and the two complementary subsets of the original integer set $\{1,\ldots,n\}$ that give rise to this value.

getHalfSet[n_, opts___Rule]:=Module[{vars, xx, ranges, nmin, vals},
  vars = Array[xx, n];
  ranges = Map[{#, 0, 1}&, vars];
  {nmin, vals} = NMinimize[spfun[vars], ranges, opts];
  {nmin, Map[Sort, splitRange[vars/.vals]]}]

As in previous examples, we explicitly set the method so that we can more readily pass it nondefault method-specific options. Finally, we set this to run many iterations with a lot of search points. Also we turn off post-processing. Why do we care about this? Well, observe that our variables are not explicitly integer valued. We are in effect fooling `NMinimize` into doing a discrete (and in fact combinatorial) optimization problem, without explicit use of discrete variables. Hence default heuristics are likely to conclude that we should attempt a "local" optimization from the final configuration produced by the differential evolution code. This will almost always be unproductive, and can take considerable time. So we explicitly disallow it. Indeed, if we have the computation time to spend, we are better off increasing our number of generations, or the size of each generation, or both.

Timing[{min, {s1, s2}} =
  getHalfSet[100, MaxIterations → 10000,
    Method → {DifferentialEvolution, CrossProbability → .8,
    SearchPoints → 100, PostProcess → False}]]

{2134.42, {2.006223098760529*^-7,
  {{1, 2, 4, 6, 7, 11, 13, 15, 16, 17, 19, 21, 23, 25, 26, 27, 31, 34,
   37, 41, 43, 44, 45, 47, 50, 51, 52, 54, 56, 66, 67, 69, 72, 73,
   75, 77, 78, 79, 80, 86, 87, 88, 89, 90, 91, 93, 96, 97, 98, 100},
  {3, 5, 8, 9, 10, 12, 14, 18, 20, 22, 24, 28, 29, 30, 32, 33, 35, 36,
   38, 39, 40, 42, 46, 48, 49, 53, 55, 57, 58, 59, 60, 61, 62, 63, 64,
   65, 68, 70, 71, 74, 76, 81, 82, 83, 84, 85, 92, 94, 95, 99}}}}

We obtain a fairly small value for our objective function. I do not know if this in fact the global minimum, and the interested reader might wish to take up this problem with an eye toward obtaining a better result.

A reasonable question to ask is how would one know, or even suspect, where to set the `CrossProbability` parameter? A method I find useful is to do "tuning runs". What this means is we do several runs with a relatively small set of search points and a fairly low bound on the number of generations (the `MaxIterations` option setting, in `NMinimize`). Once we have a feel for which values seem to be giving better results, we use them in the actual run with options settings at their full values. Suffice it to say

4.4.2 Set Partitioning via Knapsack Approach

Another approach to this problem is as follows. We take the full set and pick 100 corresponding random integer values that are either 0 or 1. An element in the set is put into one or the other subset according to the value of the bit corresponding to that element. For this to give an even split we also must impose a constraint that the size of each subset is half the total size. To get an idea of what these constraints are, we show again on our small example of size six.

vars = Array[x, 6];
ranges = Map[(0<=#<=1)&, vars];
Join[ranges, {Element[vars, Integers], Apply[Plus, vars] == 3}]

$\{0 \leq x[1] \leq 1, 0 \leq x[2] \leq 1, 0 \leq x[3] \leq 1, 0 \leq x[4] \leq 1,$
$\quad 0 \leq x[5] \leq 1, 0 \leq x[6] \leq 1, (x[1]|x[2]|x[3]|x[4]|x[5]|x[6]) \in \text{Integers},$
$\quad x[1] + x[2] + x[3] + x[4] + x[5] + x[6] == 3\}$

We are now ready to define our new objective function.

---

*Outline of **spfun2***

1. Input: a vector of integers, of even length n. All entries are 0 or 1.
2. Convert every 0 to -1.
3. Form a list of square roots of the integers in $\{1, \ldots, n\}$.
4. Multiply, componentwise, with the list of ones and negative ones.
5. Return the absolute value of the sum from step (4).

---

spfun2[vec : {__Integer}]:=Abs[(2 * vec − 1).Sqrt[N[Range[Length[vec]]]]]

Again we use our small example. What would our objective function be if the vector has ones in the first two and last places, and zeros in the middle three? First we find the exact value.

exactval = Abs[Total[Sqrt[smallset[[{1, 6, 3}]]]] − Total[Sqrt[smallset[[{5, 4, 2}]]]]]

$1 + \sqrt{2} - \sqrt{3} + \sqrt{5} - \sqrt{6}$

N[exactval]

0.468741

We see that, as expected, this agrees with our objective function.

spfun2[{1,0,1,0,0,1}]

0.468741

With this knowledge it is now reasonably straightforward to write the code that will perform our optimization. We create a set of variables, one for each element in the set. We constrain the variables to take on values that are either 0 or 1, and such that the sum is exactly half the cardinality of the set (that is, 100/2, or 50, in the example of interest to us). Since we force variables to be integer valued, NMinimize will automatically use DifferentialEvolution for its method. Again, we might still wish to explicitly request it so that we can set option to nondefault values.

Outline of getHalfSet2

1. Input: An even integer n, and options to pass along to NMinimize.
2. Create a list of variables, vars, of length n.
3. Set up constraints.
   All variables lie between 0 and 1.
   All variables are integers.
   Their total is $\frac{n}{2}$.
4. Call NMinimize, passing it spfun2[vars] as objective function, along with the constraints and the option settings that were input.
5. Return the minimum value found, and the two complementary subsets of the original integer set $\{1,\ldots,n\}$ that give rise to this value.

---

getHalfSet2[n_, opts___]:=Module[
  {vars,x,nmin,vals,ranges,s1},
  vars = Array[x,n];
  ranges = Map[(0 ≤ # ≤ 1)&, vars];
  {nmin, vals} =
   NMinimize[{spfun2[vars],
    Join[ranges, {Element[vars, Integers], Total[vars] == n/2}]}, vars, opts];
  s1 = Select[Inner[Times, Range[n], (vars/.vals), List], # ≠ 0&];
  {nmin, {s1, Complement[Range[n], s1]}}]

Timing[
  {min, {s1, s2}} = getHalfSet2[100, MaxIterations → 1000, Method →
   {DifferentialEvolution, CrossProbability → .8, SearchPoints → 100}]]

{1732.97, {0.000251303,
  {1,4,5,7,12,13,14,15,16,19,20,22,23,31,32,36,37,38,41,42,
   43,44,45,46,47,49,50,51,52,55,59,60,62,65,66,71,73,78,
   79,83,84,87,88,89,90,91,94,97,99,100},

{2,3,6,8,9,10,11,17,18,21,24,25,26,27,28,29,30,33,34,35,
39,40,48,53,54,56,57,58,61,63,64,67,68,69,70,72,74,75,
76,77,80,81,82,85,86,92,93,95,96,98}}}}

One unfamiliar with the subject might well ask what this has to do with knapsacks. The gist is as follows. A *Knapsack Problem* involves taking, or not taking, an element from a given set, and attempting to optimize some condition that is a function of those elements taken. There is a large body of literature devoted to such problems, as they subsume the *Integer Linear Programming Problem* (in short, linear program, but with variables constrained to be integer valued). It is a pleasant quality of Differential Evolution that it can be adapted to such problems.

4.4.3 Discussion of the Two Methods

The second method we showed is a classical approach in integer linear programming. One uses a set of variables constrained to be either 0 or 1 (that is, *binary* variables). We constrain their sum so that we achieve a particular goal, in this case it is that exactly half be put into one of the two subsets. While not quite a relative position indexing method, it is similar in that positions of zeros or ones determine which of two complementary subsets receives elements of the parent set.

The first method, which seemed to work better for Differential Evolution (at least with parameter settings we utilized) is less common. It is a bit mysterious, in that we use the ordering of an ensemble of reals to determine placement of individual elements of a set. This implies a certain *nonlocality* in that a change to one value can have a big effect on the interpretation of other entries. This is because it is their overall sorted ordering, and not individual values, that gets used by the objective function. Though it is not obvious that this would be useful, we saw in this example that we can get a reasonably good result.

4.5 Minimal Covering of a Set by Subsets

The problem below was once posed in the Usenet news group comp.soft-sys.math.mathematica. It is an archetypical example of the classical *subset covering problem*. In this example we are given a set of sets, each containing integers between 1 and 64. Their union is the set of all integers in that range, and we want to find a set of 12 subsets that covers that entire range. In general we would want to find a set of subsets of minimal cardinality; this is an instance where we know in advance that that cardinality is 12.

subsets = {{1,2,4,8,16,32,64},{2,1,3,7,15,31,63},{3,4,2,6,14,30,62},
{4,3,1,5,13,29,61},{5,6,8,4,12,28,60},{6,5,7,3,11,27,59},
{7,8,6,2,10,26,58},{8,7,5,1,9,25,57},{9,10,12,16,8,24,56},
{10,9,11,15,7,23,55},{11,12,10,14,6,22,54},{12,11,9,13,5,21,53},
{13,14,16,12,4,20,52},{14,13,15,11,3,19,51},{15,16,14,10,2,18,50},
{16,15,13,9,1,17,49},{17,18,20,24,32,16,48},{18,17,19,23,31,15,47},
{19,20,18,22,30,14,46},{20,19,17,21,29,13,45},{21,22,24,20,28,12,44},

{22,21,23,19,27,11,43}, {23,24,22,18,26,10,42}, {24,23,21,17,25,9,41},
{25,26,28,32,24,8,40}, {26,25,27,31,23,7,39}, {27,28,26,30,22,6,38},
{28,27,25,29,21,5,37}, {29,30,32,28,20,4,36}, {30,29,31,27,19,3,35},
{31,32,30,26,18,2,34}, {32,31,29,25,17,1,33}, {33,34,36,40,48,64,32},
{34,33,35,39,47,63,31}, {35,36,34,38,46,62,30}, {36,35,33,37,45,61,29},
{37,38,40,36,44,60,28}, {38,37,39,35,43,59,27}, {39,40,38,34,42,58,26},
{40,39,37,33,41,57,25}, {41,42,44,48,40,56,24}, {42,41,43,47,39,55,23},
{43,44,42,46,38,54,22}, {44,43,41,45,37,53,21}, {45,46,48,44,36,52,20},
{46,45,47,43,35,51,19}, {47,48,46,42,34,50,18}, {48,47,45,41,33,49,17},
{49,50,52,56,64,48,16}, {50,49,51,55,63,47,15}, {51,52,50,54,62,46,14},
{52,51,49,53,61,45,13}, {53,54,56,52,60,44,12}, {54,53,55,51,59,43,11},
{55,56,54,50,58,42,10}, {56,55,53,49,57,41,9}, {57,58,60,64,56,40,8},
{58,57,59,63,55,39,7}, {59,60,58,62,54,38,6}, {60,59,57,61,53,37,5},
{61,62,64,60,52,36,4}, {62,61,63,59,51,35,3}, {63,64,62,58,50,34,2},
{64,63,61,57,49,33,1}};

We do a brief check that the union of the subset elements is indeed the set of integers from 1 through 64.

Union[Flatten[subsets]] == Range[64]

True

4.5.1 An Ad Hoc Approach to Subset Covering

We will set up our objective function as follows. We represent a set of 12 subsets of this master set by a set of 12 integers in the range from 1 to the number of subsets (which in this example is, coincidently, also 64). This set is allowed to contain repetitions. Our objective function to minimize will be based on how many elements from 1 through 64 are "covered". Specifically it will be 2 raised to the #(elements not covered) power. The code below does this.

Outline of scfun

1. Input: a vector V of integers, a set S of subsets, and an integer n to denote the range of integers $\{1,\ldots,n\}$.
2. Compute U, the union of elements contained in the subsets S_j, for all $j \in V$.
3. Calculate c, the cardinality of the complement of our initial range by U.
   More succinctly this is $|\{1,\ldots,n\} - U|$, where subtraction is taken to mean *set complement*, and $|S|$ denotes the cardinality of S.
4. Return 2^c.

---

scfun[n : {__Integer}, set_, mx_Integer]:=
  2^Length[Complement[Range[mx], Union[Flatten[set[[n]]]]]]

This may be a bit elusive. We will examine its behavior on a specific set of subsets. Suppose we take the first 12 of our subsets.

first12 = Take[subsets, 12]

{{1,2,4,8,16,32,64}, {2,1,3,7,15,31,63}, {3,4,2,6,14,30,62},
{4,3,1,5,13,29,61}, {5,6,8,4,12,28,60}, {6,5,7,3,11,27,59},
{7,8,6,2,10,26,58}, {8,7,5,1,9,25,57}, {9,10,12,16,8,24,56},
{10,9,11,15,7,23,55}, {11,12,10,14,6,22,54}, {12,11,9,13,5,21,53}}

Their union is

elementsinfirst12 = Union[Flatten[first12]]

{1,2,3,4,5,6,7,8,9,10,11,12,13,14,15,16,21,22,23,24,25,26,27,28,
29,30,31,32,53,54,55,56,57,58,59,60,61,62,63,64}

Our objective function for this set of subsets raises 2 to the power that is the cardinality of the set of integers 1 through 64 complemented by this set. So how many elements does this union miss?

missed = Complement[Range[64], elementsinfirst12]

{17,18,19,20,33,34,35,36,37,38,39,40,41,42,43,44,
45,46,47,48,49,50,51,52}

Length[missed]

24

2^Length[missed]

16777216

Does this agree with the function we defined above? Indeed it does.

scfun[Range[12], subsets, 64]

16777216

We now give outline and code to find a set of spanning subsets.

Outline of spanningSets

1. Input: a set S of m subsets, an integer k specifying how many we are to use for our cover, and option values to pass to `NMinimize`. We assume the union of all subsets covers some range $\{1,\ldots,n\}$.
2. Create a vector of k variables.

3. Set up constraints.
    All variables are between 1 and m.
    All variables are integer valued.
4. Call NMinimize, using the constraints and scfun as defined above, along with option settings.
5. Return the minimal value (which we want to be 1, in order that there be full coverage), and the list of positions denoting which subsets we used in the cover.

spanningSets[set_, nsets_, iter_, sp_, cp_]:=Module[
  {vars, rnges, max = Length[set], nmin, vals},
  vars = Array[xx, nsets];
  rnges = Map[(1 ≤ # ≤ max)&, vars];
  {nmin, vals} = NMinimize[
    {scfun[vars, set, max], Append[rnges, Element[vars, Integers]]},
    vars, MaxIterations → iter,
    Method → {DifferentialEvolution, SearchPoints → sp, CrossProbability → cp}];
  vals = Union[vars/.vals];
  {nmin, vals}]

In small tuning runs I found that a fairly high crossover probability setting seemed to work well.

Timing[{min, sets} = spanningSets[subsets, 12, 700, 200, .94]]

{365.099, {1., {1, 7, 14, 21, 24, 28, 34, 35, 47, 52, 54, 57}}}

Length[Union[Flatten[subsets[[sets]]]]]

64

While this is not lightning fast, we do obtain a good result in a few minutes of run time.

We note that while this was not coded explicitly to use relative position indexing, it could have been. That is, we could have used vectors of 64 values between 0 and 1, and taken the positions of the smallest 12 to give 12 members of subsets. The interested reader may wish to code this variant.

4.5.2 Subset Covering via Knapsack Formulation

Another method is to cast this as a standard knapsack problem. First we transform each of our set of subsets into a *bit vector* representation. In this form each subset is represented by a positional list of zeros and ones. In effect we are translating from a *sparse* to a *dense* representation.

4 Relative Position Indexing Approach

---

*Outline of **densevec***

1. Input: a set S of integers and a length n. It is assumed that the members of S all lie in $\{1,\ldots,n\}$.
2. Create a vector V of length n. Initialize all elements to be 0.
3. Loop: For each $j \in S$, set the jth element of V to be 1.
4. Return V.

---

densevec[spvec_, len_]:=Module[
  {vec = Table[0, {len}]},
  Do[vec[[spvec[[j]]]] = 1, {j, Length[spvec]}];
  vec]

We now apply this function to each member of our set of subsets, that is, make a dense representation of each subset.

mat = Map[densevec[#, 64]&, subsets];

It might not be obvious what we have done, so we illustrate using the fourth of our 64 matrix rows.

mat[[4]]

$\{1,0,1,1,1,0,0,0,0,0,0,0,1,0,0,0,0,0,0,0,0,0,0,0,0,0,0,0,1,0,0,0,0,$
$0,0,0,0,0,0,0,0,0,0,0,0,0,0,0,0,0,0,0,0,0,0,0,0,0,0,0,1,0,0,0\}$

We have ones at positions that correspond to the elements contained in our fourth subset, and zeros elsewhere. Specifically, the ones are at the positions shown below.

Flatten[Position[mat[[4]], 1]]

$\{1,3,4,5,13,29,61\}$

But this is, up to ordering, exactly the elements in the fourth subset. That is, we pass a basic consistency check.

Sort[subsets[[4]]]

$\{1,3,4,5,13,29,61\}$

As in our last knapsack problem, we again work with binary variables and minimize their sum, subject to certain constraints. We use a binary variables to represent each subset. A one means we use that subset in our set cover, and a zero means we do not.

Let us consider the vector of those zeros and ones. Now our requirement is that we fully cover the superset, that is, the range of integers from 1 to 64.

How might we impose this? Well, let us take a look at the dot product of such a vector with the matrix of bit vectors that we already formed. Again we use as an example the first 12 subsets, so our vector representing this has ones in the first 12 slots, and zeros in the remaining 64-12=52 slots.

first12vec = Join[ConstantArray[1, 12], ConstantArray[0, 52]]

{1,1,1,1,1,1,1,1,1,1,1,1,0,0,0,0,0,0,0,0,0,0,0,0,0,0,0,0,0,0,0,0,
 0,0,0,0,0,0,0,0,0,0,0,0,0,0,0,0,0,0,0,0,0,0,0,0,0,0,0,0,0,0,0,0}

What does the dot product of this vector with our matrix of bit vectors represent? Well, let's consider the meaning of this matrix for a moment. A one in row j, column k means that subset j contains element k. One then realizes that the dot product gives us the following information. The kth element in the result will be a nonnegative number (possibly zero), and represent the number of times that k appears in the union of subsets represented by `first12vec`. So the condition we will need to impose in our optimization is that the dot product of this vector with our matrix of bitvectors has all positive entries. Notice that `first12vec` fails to satisfy this condition.

first12vec.mat

{4,4,4,4,5,5,5,5,4,4,4,4,2,2,2,2,0,0,0,0,1,1,1,1,1,1,1,1,1,1,1,0,
 0,0,0,0,0,0,0,0,0,0,0,0,0,0,0,0,0,1,1,1,1,1,1,1,1,1,1,1,1,1,1,1}

---

*Outline of **spanningSets2***

1. Input: a set S of m subsets (each now represented as a bit vector), and option values to pass to `NMinimize`. We assume the union of all subsets covers some range $\{1,\ldots,n\}$.
2. Create a vector of m variables, `vars`.
3. Set up constraints.
    All variables lie between 0 and 1.
    All variables are integer valued.
    The union of subsets corresponding to variables with value of 1 covers the full range $\{1,\ldots,n\}$. This is done by checking that each elements of S.vars is greater or equal to 1.
4. Call `NMinimize` to minimize the sum of vars, subject to the above constraints, using the input option settings.
5. Return the minimal value and the list of positions denoting which subsets we used in the cover.

```
spanningSets2[set_, iter_, sp_, seed_, cp_:.5]:=Module[
  {vars, rnges, max = Length[set], nmin, vals},
  vars = Array[xx, max];
  rnges = Map[(0 ≤ # ≤ 1)&, vars];
  {nmin, vals} =
  NMinimize[{Apply[Plus, vars], Join[rnges, {Element[vars, Integers]},
    Thread[vars.set ≥ Table[1, {max}]]]}, vars, MaxIterations → iter,
    Method → {DifferentialEvolution, CrossProbability → cp,
    SearchPoints → sp, RandomSeed → seed}];
  vals = vars/.vals;
  {nmin, vals}]

Timing[{min, sets} = spanningSets2[mat, 2000, 100, 0, .9]]

{1930.4Second, {12., {0,0,1,0,0,1,0,0,0,0,0,0,0,0,0,1,0,0,
  0,0,1,0,0,0,1,0,0,0,0,1,0,0,0,0,0,0,1,0,0,0,0,1,0,
  0,0,0,1,0,0,0,0,1,0,0,1,0,0,0,0,0,0,0,0,1}}}
```

We have again obtained a result that uses 12 subsets. We check that it covers the entire range.

Total[Map[Min[#, 1]&, sets.mat]]

64

We see that this method was much slower. Experience indicates that it needs a lot of iterations and careful setting of the CrossProbability option. So at present `NMinimize` has difficulties with this formulation. All the same it is encouraging to realize that one may readily set this up as a standard knapsack problem, and still hope to solve it using Differential Evolution. Moreover, as the alert reader may have observed, we actually had an added benefit from using this method: nowhere did we need to assume that minimal coverings require 12 subsets.

4.6 An Assignment Problem

Our next example is a benchmark from the literature of discrete optimization. We are given two square matrices. We want a permutation that, when applied to the rows and columns of the second matrix, multiplied element-wise with corresponding elements of the first, and all elements summed, gives a minimum value. The matrices we use have 25 rows. This particular example is known as the NUG25 problem. It is an example of a *Quadratic Assignment Problem* (QAP). The optimal result is known and was verified by a large parallel computation. We mention that the methods of handling this problem can, with minor modification, be applied to related problems that require the selecting of a permutation (for example, the traveling salesman problem).

mat1 = {{0,1,2,3,4,1,2,3,4,5,2,3,4,5,6,3,4,5,6,7,4,5,6,7,8},
{1,0,1,2,3,2,1,2,3,4,3,2,3,4,5,4,3,4,5,6,5,4,5,6,7},
{2,1,0,1,2,3,2,1,2,3,4,3,2,3,4,5,4,3,4,5,6,5,4,5,6},
{3,2,1,0,1,4,3,2,1,2,5,4,3,2,3,6,5,4,3,4,7,6,5,4,5},
{4,3,2,1,0,5,4,3,2,1,6,5,4,3,2,7,6,5,4,3,8,7,6,5,4},
{1,2,3,4,5,0,1,2,3,4,1,2,3,4,5,2,3,4,5,6,3,4,5,6,7},
{2,1,2,3,4,1,0,1,2,3,2,1,2,3,4,3,2,3,4,5,4,3,4,5,6},
{3,2,1,2,3,2,1,0,1,2,3,2,1,2,3,4,3,2,3,4,5,4,3,4,5},
{4,3,2,1,2,3,2,1,0,1,4,3,2,1,2,5,4,3,2,3,6,5,4,3,4},
{5,4,3,2,1,4,3,2,1,0,5,4,3,2,1,6,5,4,3,2,7,6,5,4,3},
{2,3,4,5,6,1,2,3,4,5,0,1,2,3,4,1,2,3,4,5,2,3,4,5,6},
{3,2,3,4,5,2,1,2,3,4,1,0,1,2,3,2,1,2,3,4,3,2,3,4,5},
{4,3,2,3,4,3,2,1,2,3,2,1,0,1,2,3,2,1,2,3,4,3,2,3,4},
{5,4,3,2,3,4,3,2,1,2,3,2,1,0,1,4,3,2,1,2,5,4,3,2,3},
{6,5,4,3,2,5,4,3,2,1,4,3,2,1,0,5,4,3,2,1,6,5,4,3,2},
{3,4,5,6,7,2,3,4,5,6,1,2,3,4,5,0,1,2,3,4,1,2,3,4,5},
{4,3,4,5,6,3,2,3,4,5,2,1,2,3,4,1,0,1,2,3,2,1,2,3,4},
{5,4,3,4,5,4,3,2,3,4,3,2,1,2,3,2,1,0,1,2,3,2,1,2,3},
{6,5,4,3,4,5,4,3,2,3,4,3,2,1,2,3,2,1,0,1,4,3,2,1,2},
{7,6,5,4,3,6,5,4,3,2,5,4,3,2,1,4,3,2,1,0,5,4,3,2,1},
{4,5,6,7,8,3,4,5,6,7,2,3,4,5,6,1,2,3,4,5,0,1,2,3,4},
{5,4,5,6,7,4,3,4,5,6,3,2,3,4,5,2,1,2,3,4,1,0,1,2,3},
{6,5,4,5,6,5,4,3,4,5,4,3,2,3,4,3,2,1,2,3,2,1,0,1,2},
{7,6,5,4,5,6,5,4,3,4,5,4,3,2,3,4,3,2,1,2,3,2,1,0,1},
{8,7,6,5,4,7,6,5,4,3,6,5,4,3,2,5,4,3,2,1,4,3,2,1,0}};

mat2 = {{0,3,2,0,0,10,5,0,5,2,0,0,2,0,5,3,0,1,10,0,2,1,1,1,0},
{3,0,4,0,10,0,0,2,2,1,5,0,0,0,0,0,1,6,1,0,2,2,5,1,10},
{2,4,0,3,4,5,5,5,1,4,0,4,0,4,0,3,2,5,5,2,0,0,3,1,0},
{0,0,3,0,0,0,2,2,0,6,2,5,2,5,1,1,1,2,2,4,2,0,2,2,5},
{0,10,4,0,0,2,0,0,0,0,0,0,0,2,0,0,2,0,5,0,2,1,0,0,2},
{10,0,5,0,2,0,10,10,5,10,6,0,0,10,2,10,1,5,5,2,5,0,0,2,0,1},
{5,0,5,2,0,10,0,1,3,5,0,0,2,4,5,10,6,0,5,5,5,0,5,5,0},
{0,2,5,2,0,10,1,0,10,2,5,2,0,3,0,0,0,4,0,5,0,5,2,2,5},
{5,2,1,0,0,5,3,10,0,5,6,0,1,5,5,5,2,3,5,0,2,10,10,1,5},
{2,1,4,6,0,10,5,2,5,0,0,1,2,1,0,0,0,0,6,6,4,5,3,2,2},
{0,5,0,2,0,6,0,5,6,0,0,2,0,4,2,1,0,6,2,1,5,0,0,1,5},
{0,0,4,5,0,0,0,2,0,1,2,0,2,1,0,3,10,0,0,4,0,0,4,2,5},
{2,0,0,2,0,0,2,0,1,2,0,2,0,4,5,0,1,0,5,0,0,0,5,1,1},
{0,0,4,5,0,10,4,3,5,1,4,1,4,0,0,0,2,2,0,2,5,0,5,2,5},
{5,0,0,1,2,2,5,0,5,0,2,0,5,0,0,2,0,0,0,6,3,5,0,0,5},
{3,0,3,1,0,10,10,0,5,0,1,3,0,0,2,0,0,5,5,1,5,2,1,2,10},
{0,1,2,1,0,1,6,0,2,0,0,10,1,2,0,0,0,5,2,1,1,5,6,5,5},
{1,6,5,2,2,5,0,4,3,0,6,0,0,2,0,5,5,0,4,0,0,0,0,5,0},
{10,1,5,2,0,5,5,0,5,6,2,0,5,0,0,5,2,4,0,5,4,4,5,0,2},

{0,0,2,4,5,2,5,5,0,6,1,4,0,2,6,1,1,0,5,0,4,4,1,0,2},
{2,2,0,2,0,5,5,0,2,4,5,0,0,5,3,5,1,0,4,4,0,1,0,10,1},
{1,2,0,0,2,0,0,5,10,5,0,0,0,0,5,2,5,0,4,4,1,0,0,0,0},
{1,5,3,2,1,2,5,2,10,3,0,4,5,5,0,1,6,0,5,1,0,0,0,0,0},
{1,1,1,2,0,0,5,2,1,2,1,2,1,2,0,2,5,5,0,0,10,0,0,0,2},
{0,10,0,5,2,1,0,5,5,2,5,5,1,5,5,10,5,0,2,2,1,0,0,2,0}};

First we define a function to permute rows and columns of a matrix. It simply rearranges the matrix so that both rows and columns are reordered according to a given permutation.

---

*Outline of **permuteMatrix***

1. Input: a square matrix M and a permutation P of the set $\{1,\ldots,n\}$, where n is the dimension of M.
2. Form $\tilde{M}$, the matrix obtained by rearranging rows and columns of M as specified by P.
3. Return $\tilde{M}$.

---

permuteMatrix[mat_, perm_]:=mat[[perm, perm]]

We use a small matrix to see how this works.

MatrixForm[mat = Array[x, {4,4}]]

$$\begin{pmatrix} x[1,1] & x[1,2] & x[1,3] & x[1,4] \\ x[2,1] & x[2,2] & x[2,3] & x[2,4] \\ x[3,1] & x[3,2] & x[3,3] & x[3,4] \\ x[4,1] & x[4,2] & x[4,3] & x[4,4] \end{pmatrix}$$

Now we move rows/columns (4,1,3,2) to positions (1,2,3,4), and observe the result.

MatrixForm[permuteMatrix[mat, {4,1,2,3}]]

$$\begin{pmatrix} x[4,4] & x[4,1] & x[4,2] & x[4,3] \\ x[1,4] & x[1,1] & x[1,2] & x[1,3] \\ x[2,4] & x[2,1] & x[2,2] & x[2,3] \\ x[3,4] & x[3,1] & x[3,2] & x[3,3] \end{pmatrix}$$

Let us return to the NUG25 problem. Below is an optimal permutation (it is not unique). We remark that the computation that verified the optimality took substantial time and parallel resources.

$p = \{5, 11, 20, 15, 22, 2, 25, 8, 9, 1, 18, 16, 3, 6, 19, 24, 21, 14, 7, 10, 17, 12, 4, 23, 13\};$

We compute the objective function value we obtain from this permutation. As a sort of baseline, we show the result one obtains from applying no permutation. We then compute results of applying several random permutations. This gives some idea of how to gauge the results below.

best = Apply[Plus, Flatten[mat1 * permuteMatrix[mat2, p]]]

3744

baseline = Apply[Plus, Flatten[mat1 * mat2]]

4838

**randomvals = Table[
 perm = Ordering[RandomReal[{0, 1}, {25}]];
 Apply[Plus, Flatten[mat1 * permuteMatrix[mat2, perm]]], {10}]**

{4858, 5012, 5380, 5088, 4782, 4994, 5032, 5044, 5088, 5094}

A substantially longer run over random permutations gives an indication of how hard it is to get good results via a naive random search.

**SeedRandom[1111];
Timing[randomvals = Table[
 perm = Ordering[RandomReal[{0, 1}, {25}]];
 Total[Flatten[mat1 * permuteMatrix[mat2, perm]]],
 {1000000}];]**

{449.06, Null}

Min[randomvals]

4284

4.6.1 Relative Position Indexing for Permutations

We must decide how to make a set of values into a permutation. Our first approach is nearly identical to the ensemble order method we used on the set partition problem. Specifically, we will let the Ordering function of a set of real values determine a permutation.

Outline of QAP

1. Input: square matrices M_1 and M_2 each of dimension n, along with parameter settings to pass to NMinimize.
2. Form a vector of variables of length n. Give them initial ranges from 0 to 1.

3. Form an objective function that sums the n^2 products of elements of the first matrix and elements of the row-and-column permuted second matrix.

   The permutation is determined by the ordering of values of the variables vector. (Remark: some readers might recognize this as a *matrix inner product* computed via the *matrix trace* of the usual matrix product).

   For improved speed (at the cost of memory) we *memoize* values of the objective function. What that means is we record them once computed, so that recomputation is done by fast lookup. Readers familiar with data structure methods may recognize this as an application of *hashing*.
4. Call NMinimize on the objective function, using the above ranges, constraints, and input option settings.
5. Return the minimal value found, along with the permutation that gives rise to that value.

QAP[mat1_, mat2_, cp_, it_, sp_, sc_]:=Module[
{len = Length[mat1], obfunc, obfunc2, vars, x, nmin, vals, rnges},
vars = Array[x, len];
rnges = Map[{#, 0, 1}&, vars];
obfunc[vec : {_Real}]:=obfunc2[Ordering[vec]];

obfunc2[perm_]:=obfunc2[perm] =
  Total[Flatten[mat1 ∗ permuteMatrix[mat2, perm]]];
{nmin, vals} = NMinimize[obfunc[vars], rnges, MaxIterations → it,
  Method → {DifferentialEvolution, SearchPoints → sp, CrossProbability → cp,
    ScalingFactor → sc, PostProcess → False}];
Clear[obfunc2];
{nmin, Ordering[vars/.vals]}]

Again we face the issue that this problem requires nonstandard values for options to the DifferentialEvolution method, in order to achieve a reasonable result. While this is regretable it is clearly better than having no recourse at all. The idea behind having CrossProbability relatively small is that we do not want many crossovers in mating a pair of vectors. This in turn is because of the way we define a permutation. In particular it is not just values but relative values across the entire vector that give us the permutation. Thus disrupting more than a few, even when mating a pair of good vectors, is likely to give a bad vector. This was also the case with the set partitioning example we encountred earlier.

We saw that the baseline permutation (do nothing) and random permutations tend to be far from optimal, and even a large random sampling will get us only about half way from baseline to optimal. A relatively brief run with "good" values for the algorithm parameters, on the other hand, yields something notably better. (In the next subsection we explicitly show how one might use short *tuning runs* to find such parameter settings.)

SeedRandom[11111];
Timing[{min, perm} = QAP[mat1, mat2, .06, 200, 40, .6]]

{13.5048, {3864.,
 {22, 20, 17, 12, 5, 13, 15, 23, 25, 2, 19, 10, 9,
 8, 4, 1, 7, 6, 16, 18, 24, 21, 14, 3, 11}}}

We now try a longer run.

SeedRandom[11111];
Timing[{min, perm} = QAP[mat1, mat2, .06, 4000, 100, .6]]

{394.881, {3884.,
 {15, 20, 19, 10, 13, 22, 1, 16, 7, 4, 9, 25, 6, 23,
 12, 8, 11, 21, 14, 17, 5, 2, 18, 3, 24}}}

We learn a lesson here. Sometimes a short run is lucky, and a longer one does not fare as well. We will retry with a different crossover, more iterations, and a larger set of chromosomes.

Timing[{min, perm} = QAP[mat1, mat2, .11, 10000, 200, .6]]

{2186.43, {3826.,
 {5, 2, 18, 11, 4, 12, 25, 8, 14, 24, 17, 3, 16, 6, 21,
 20, 23, 9, 7, 10, 22, 15, 19, 1, 13}}}

This result is not bad.

4.6.2 Representing and Using Permutations as Shuffles

The method we now show will generate a permutation as a *shuffle* of a set of integers. We first describe a standard way to shuffle, with uniform probability, a set of n elements. First we randomly pick a number j_1 in the range $\{1, \ldots, n\}$ and, if $j_1 \neq 1$, we swap the first and j_1th elements. We then select at random an element j_2 in the range $\{2, \ldots, n\}$. If $j_2 \neq 2$ we swap the second and j_2th elements. The interested reader can convince him or herself that this indeed gives a uniform random shuffle (in contrast, selecting all elements in the range $\{1, \ldots, n\}$ fails to be uniform).

Our goal, actually, is not directly to generate shuffles, but rather to use them. Each chromosome will represent a shuffle, encoded as above by a set of swaps to perform. So the effective constraint on the first variable is that it be an integer in the range $\{1, \ldots, n\}$, while the second must be an integer in the range $\{2, \ldots, n\}$, and so on (small point: we do not actually require an nth variable, since its value must always be n). We require a utility routine to convert quickly from a shuffle encoding to a simple permutation vector. The code below will do this. We use the `Compile` function of *Mathematica* to get a speed boost.

Outline of **getPerm**

1. Input: a shuffle S encoded as $n-1$ integers in the range $\{1,\ldots,n\}$, with the jth actually restricted to lie in the subrange $\{j,\ldots,n\}$.
2. Initialize a vector P of length n to be the identity permutation (that is, the ordered list $\{1,\ldots,n\}$).
3. Iterate over S.
4. Swap the jth element of P with the element whose index is the (current) jth element of S.
5. Return P.

getPerm = Compile[{{shuffle, _Integer, 1}}, Module[
  {perm, len = Length[shuffle] + 1},
  perm = Range[len];
  Do[perm[[{j, shuffle[[j]]}]] = perm[[{shuffle[[j]], j}]], {j, len − 1}];
  perm]];

Okay, maybe that was a bit cryptic. Here is a brief example that will shed light on this process. Say our shuffle encoding for a set of five elements is $\{2,4,5,4\}$. What would this do to permute the set $\{1,2,3,4,5\}$? First we swap elements 1 and 2, so we have $\{2,1,3,4,5\}$. We next swap elements 2 and 4, giving $\{2,4,3,1,5\}$. Then we swap elements 3 and 5 to obtain $\{2,4,5,1,3\}$. Finally, as the fourth element in our shuffle is a 4, we do no swap. Let us check that we did indeed get the permutation we claim.

getPerm[{2, 4, 5, 4}]

$\{2,4,5,1,3\}$

The constraints we would like to enforce are that all chromosome elements be integers, and that the jth such element be between j and the total length inclusive. The bit of code below will show how we might set up such constraints.

len = 5;
vars = Array[x, len − 1];
constraints = Prepend[Map[(#[[1]] ≤ # ≤ len)&, vars], Element[vars, Integers]]

$\{(x[1]|x[2]|x[3]|x[4]) \in \text{Integers}, 1 \le x[1] \le 5, 2 \le x[2] \le 5, 3 \le x[3] \le 5, 4 \le x[4] \le 5\}$

There is a small wrinkle. It is often faster not to insist on integrality, but rather to use real numbers and simply round off (or truncate). To get uniform probabilities initially, using rounding, we constrain so that a given variable is at least its minimal allowed integer value minus 1/2, and at most its maximal integer value plus 1/2.

Without further fuss, we give an outline and code for this optimization approach.

Outline of QAP2

1. Input: square matrices M_1 and M_2 each of dimension n, along with parameter settings to pass to NMinimize.
2. Form a vector of variables of length $n-1$. For j in $\{1,\ldots,n-1\}$ constrain the jth variable to lie in the range $\{j-.499\ldots,n+1.499\}$.
3. Form an objective function that sums the n^2 products of elements of the first matrix and elements of the row-and-column permuted second matrix. The variables vector, with entries rounded to nearest integers, may be viewed as a shuffle on a set of n elements. The permutation is determined by invoking getPerm on the variables vector.
4. Call NMinimize on the objective function, using the above variables, constraints, and input option settings.
5. Return the minimal value found, along with the permutation that gives rise to that value.

QAP2[mat1_, mat2_, cp_, it_, sp_]:=Module[
{len = Length[mat1] − 1, obfunc, vars, x, nmin, vals, constraints},
vars = Array[x, len];
constraints = Map[(#[[1]] − .499 ≤ # ≤ len + 1.499)&, vars];
obfunc[vec : {_Real}]:=
 Total[Flatten[mat1 ∗ permuteMatrix[mat2, getPerm[Round[vec]]]]];
{nmin, vals} = NMinimize[{obfunc[vars], constraints}, vars,
 Method → {DifferentialEvolution, SearchPoints → sp, CrossProbability → cp,
 PostProcess → False}, MaxIterations → it, Compiled → False];
{nmin, getPerm[Round[vars/.vals]]}]

We show a sample tuning run. We keep the number of iterations and number of chromosomes modest, and try cross probabilities between 0.05 and 0.95, at increments of .05.

Quiet[Table[{j, First[QAP2[mat1, mat2, j/100, 50, 20]]}, {j, 5, 95, 5}]]

{{5, 4364.}, {10, 4436.}, {15, 4538.}, {20, 4428.}, {25, 4522.},
 {30, 4506.}, {35, 4518.}, {40, 4550.}, {45, 4512.}, {50, 4456.},
 {55, 4530.}, {60, 4474.}, {65, 4520.}, {70, 4412.}, {75, 4474.},
 {80, 4454.}, {85, 4410.}, {90, 4314.}, {95, 4324.}}

From this we home in on the region of the larger values since they seem to be consistently a bit better than other values (it is interesting that this is the opposite of what I had found for the relative index positioning approach in the previous subsection). We now do larger runs to get a better idea of what are the relative merits of these various cross probability parameter settings.

Quiet[Table[{ j, First[QAP2[mat1, mat2, $j/100, 80, 20$]]}, { $j, 87, 98, 1$}]]

{{87, 4298.}, {88, 4418.}, {89, 4346.}, {90, 4314.}, {91, 4396.}, {92, 4416.},
{93, 4300.}, {94, 4308.}, {95, 4274.}, {96, 4322.}, {97, 4282.}, {98, 4298.}}

We will finally try a longer run with cross probability set to 0.975.

Quiet[Timing[{min, perm} = QAP2[mat1, mat2, .975, 10000, 100]]]

{2590.27, {3814.,
  {5, 2, 11, 22, 15, 18, 25, 16, 9, 1, 17, 3, 6, 8,
   19, 12, 14, 7, 23, 20, 24, 4, 21, 10, 13}}}

This gets us reasonably close to the global minimum with a scant 15 lines of code. While it is mildly more complicated than the 10 line relative position indexing method, it has the advantage that it is slightly less dependent on fine tuning of the cross probability parameter.

4.6.3 Another Shuffle Method

There are other plausible ways to set up permutations, such that they behave in a reasonable manner with6 respect to mutation and mating operations. Here is one such.

We have for our vector a set of integers from 1 to n, the length of the set in question (again we will actually work with reals, and round off to get integers). The range restriction is the only stipulation and in particular it may contain repeats. We associate to it a unique permutation as follows. We initialize a list to contain n zeros. The first element in our list is then set to the first element in the vector. We also have a marker set telling us that that first element is now used. We iterate over subsequent elements in our list, setting them to the corresponding values in vector provided those values are not yet used. Once done with this iteration we go through the elements that have no values, assigning them in sequence the values that have not yet been assigned. This method, which is used in [7], is similar to that of `GeneRepair` [8]. It is also related to a method of [12], although they explicitly alter the recombination (that is, the genotype) rather than the resulting phenotype.

---

*Outline of **getPerm2***

1. Input: a shuffle S encoded as n integers in the range $\{1,\ldots,n\}$.
2. Create vectors P_1 and P_2 of length n. The first will be for the permutation we create, and the second will mark as "used" those elements we have encountered. Initialize elements of each to be 0.

3. Loop over S. Denote by k the jth element of S. If the kth element of P_2 is 0, this means we have not yet used k in our permutation.
   Set $P_2(k)$ to j to mark it as used.
   Set $P_1(j)$ to k.
4. Initialize a counter k to 1.
5. Loop over P_1. If the jth element, $P_1(j)$, is 0 then it needs to be filled in with a positive integer not yet used.
   Find smallest k for which $P_2(k)$ is 0 (telling us that k is not used as yet in the permutation).
   For that k, set $P_1(j)$ to be k, and mark $P_2(k)$ nonzero (alternatively, could simply increment k so it will not revisit this value).
6. Return P_1.

```
getPerm2 = Compile[{{vec, _Integer, 1}}, Module[
  {p1, p2, len = Length[vec], k}, p1 = p2 = Table[0, {len}];
  Do[k = vec[[j]];
    If[p2[[k]] == 0, p2[[k]] = j; p1[[j]] = k;], {j, len}];
  k = 1;
  Do[If[p1[[j]] == 0, While[p2[[k]] ≠ 0, k++];
    p1[[j]] = k;
    p2[[k]] = j], {j, len}];
  p1]];
```

We illustrate with a small example. Say we have the vector $\{4, 1, 4, 3, 1\}$. What permutation does this represent? Well, we have a 4 in the first slot, so the resulting permutation vector starts with 4. Then we have a 1, so that's the next element in the permutation. Next is a 4, which we have already used. We defer on that slot. Next is a 3, so the fourth slot in our permutation is 3. last is a 1, which we have already encountered, so we defer on filling in the fifth position of our permutation. We have completed one pass through the permutation. The entries we were unable to use were in positions 3 and 5. The values not yet used are 2 and 5 (because we filled in a vector as $\{4, 1, x, 3, y\}$, where x and y are not yet known). We now simply use these in order, in the empty slots. That is, entry 3 is 2 and entry 5 is 5. We obtain as our permutation $\{4, 1, 2, 3, 5\}$.

getPerm2[{4,1,4,3,1}]

$\{4, 1, 2, 3, 5\}$

This notion of associating a list with repeats to a distinct shuffle has a clear drawback insofar as earlier elements are more likely than later ones to be assigned to their corresponding values in the vector. All the same, this provides a reasonable way to make a chromosome vector containing repeats correspond to a permutation (and once the method has started to produce permutations, mating/mutation will not cause too many repeats provided the crossover probability is either fairly low or fairly high). Moreover, one can see that any sensible mating process of two chromosomes will less drastically alter the objective function than would be the case in the ensemble ordering, as the corresponding permutation now depends far less on overall ordering in the chromosomes. The advantage is that this method will thus be somewhat less in need of intricate tuning for the crossover probability parameter (but we will do that anyway).

Outline of **QAP3**

1. Input: square matrices M_1 and M_2 each of dimension n, along with parameter settings to pass to `NMinimize`.
2. Form a vector of variables of length n. Constrain each variable to lie in the range $\{.501\ldots, n+.499\}$.
3. Form an objective function that sums the n^2 products of elements of the first matrix and elements of the row-and-column permuted second matrix.

   The variables vector, with entries rounded to nearest integers, may be viewed as a shuffle on a set of n elements. The permutation is determined by invoking `getPerm2` on the variables vector.
4. Call `NMinimize` on the objective function, using the above variables, constraints, and input option settings.
5. Return the minimal value found, along with the permutation that gives rise to that value.

QAP3[mat1_, mat2_, cp_, it_, sp_]:=Module[
 {len = Length[mat1], obfunc, vars, x, nmin, vals, constraints},
 vars = Array[x, len];
 constraints = Map[(.501 ≤ # ≤ len + 0.499)&, vars];
 obfunc[vec : {_Real}]:=
  Total[Flatten[mat1 * permuteMatrix[mat2, getPerm2[Round[vec]]]]];
 {nmin, vals} = NMinimize[{obfunc[vars], constraints}, vars,
   Method → {DifferentialEvolution, SearchPoints → sp,
    CrossProbability → cp, PostProcess → False},
   MaxIterations → it, Compiled → False];
 {nmin, getPerm2[Round[vars/.vals]]}]

We'll start with a tuning run.

Quiet[Table[{j, First[QAP3[mat1, mat2, j/100, 50, 20]]}, {j, 5, 95, 5}]]

{{5, 4486.}, {10, 4498.}, {15, 4464.}, {20, 4492.}, {25, 4430.},
 {30, 4516.}, {35, 4482.}, {40, 4396.}, {45, 4432.}, {50, 4472.},
 {55, 4548.}, {60, 4370.}, {65, 4460.}, {70, 4562.}, {75, 4398.},
 {80, 4466.}, {85, 4378.}, {90, 4426.}, {95, 4354.}}

I did a second run (not shown), in the upper range of crossover probabilities, and with more iterations and larger numbers of search points. It homed in on .93 as a reasonably good choice for a crossover probability setting.

Timing[Quiet[QAP3[mat1, mat2, .93, 8000, 100]]]

{2380.2, {3888.,
  {7, 20, 11, 8, 13, 4, 25, 10, 19, 18, 17, 22, 6, 3, 5, 15, 24,
  14, 23, 21, 1, 16, 2, 12, 9, 26}}}

4.7 Hybridizing Differential Evolution for the Assignment Problem

Thus far we have seen methods that, for a standard benchmark problem from the quadratic assignment literature, take us to within shouting distance of the optimal value. These methods used simple tactics to formulate permutations from a vector chromosome, and hence could be applied within the framework of Differential Evolution. We now show a method that hybridizes Differential Evolution with another approach.

A common approach to combinatorial permutation problems is to swap pairs (this is often called *2-opt*), or reorder triples, of elements (also reversal of segments is common). With Differential Evolution one might do these by modifying the objective function to try them, and then recording the new vector (if we choose to use it) in the internals of the algorithm. This can be done in `NMinimize`, albeit via alteration of an entirely undocumented internal variable. We show this below, using a simple set of pair swaps. When we obtain improvement in this fashion, we have gained something akin to a local *hill climbing* method. I remark that such hybridization, of an evolutionary method with a local improvement scheme, is often referred to as a *memetic* algorithm. Nice expositions of such approaches can be found in [9] and [6].

The code creates a random value to decide when to use a swap even if it resulted in no improvement. This can be a useful way to maintain variation in the chromosome set. We also use a print flag: if set to `True`, whenever we get an improvement on the current best permutation, we learn what is the new value and how much time elapsed since the last such improvement. We also learn when we get such an improvement arising from a local change (that is, a swap).

As an aside, the use of a swap even when it gives a worse result has long standing justification. The idea is that we allow a decrease in quality in the hope that it will later help in finding an improvement. This is quite similar to the method of *simulated annealing*, except we do not decrease the probability, over the course of generations, of accepting a decrease in quality.

---

*Outline of **QAP4***

1. Input: square matrices M_1 and M_2 each of dimension n, along with parameter settings to pass to `NMinimize`, and a probability level p between 0 and 1 to determine when to retain an altered chromosome that gives a decrease in quality.
2. Form a vector of variables of length n.

3. Give them initial ranges from 0 to 1.
4. Form an objective function that sums the n^2 products of elements of the first matrix and elements of the row-and-column permuted second matrix. As in QAP, the permutation is determined by the ordering of values of the variables vector.
5. Iterate some number of times (a reasonable value is 4).
   Swap a random pair of elements in the variables vector.
   Check whether we got improvement in the objective function.
   If so, keep this improved vector.
   If not, possibly still keep it depending on whether a random value between 0 and 1 is larger than p, and also whether the better vector is the best seen thus far (we never replace the best one we have).
   Depending on an input flag setting, either restart the swapping (if we are not done iterating) with our original vector, or else continue with the one created from prior swaps.
6. Call NMinimize on the objective function, using the above ranges, constraints, and input option settings.
7. Return the minimal value found, along with the permutation that gives rise to that value.

```
QAP4[mat1_, mat2_, cp_, it_, sp_, sc_, maxj_:4, keep_:0.4, restorevector_,
printFlag_:False]:=Module[
{len = Length[mat1], objfunc, objfunc2, vars, vv, nmin, vals, rnges, best,
  bestvec, indx = 0, i = 0, tt = TimeUsed[]},
vars = Array[vv, len];
rnges = Map[{#, 0, 1}&, vars];
objfunc2[vec_]:=objfunc2[vec] =
Total[Flatten[mat1 * permuteMatrix[mat2, vec]]];
objfunc[vec : {_Real}]:=Module[
  {val1, val2, r1, r2, vec1 = vec, vec2 = vec, max = Max[Abs[vec]], j = 0},
  {vec1, vec2} = {vec1, vec2}/max;
  val1 = objfunc2[Ordering[vec1]];
  While[j ≤ maxj,
  j++;
  {r1, r2} = RandomInteger[{1, len}, {2}];
  If[restorevector, vec2 = vec1];
  vec2[[{r1, r2}]] = vec2[[{r2, r1}]];
  val2 = objfunc2[Ordering[vec2]];
  If[val2 < best, j--;
    If[printFlag, Print["locally improved", {best, val2}]]];
    If[val2 ≤ val1||(val1 > best&&RandomReal[] > keep),
      OptimizeNMinimizeDumpvec = vec2;
      If[val2 < val1, vec1 = vec2];
```

```
    val1 = Min[val1, val2],
    OptimizeNMinimizeDump`vec = vec1];
  If[val1 < best,
    best = val1;
    vec1 = bestvec = OptimizeNMinimizeDump`vec;
    If[printFlag,
      Print["new low ", ++indx, " {iteration, elapsedtime, newvalue} ",
      {i, TimeUsed[] − tt, best}]]; tt = TimeUsed[];];
    ];
    val1];
  bestvec = Range[len];
  best = Total[Flatten[mat1 ∗ mat2]];
  {nmin, vals} = NMinimize[objfunc[vars], rnges,
    MaxIterations → it, Compiled → False, StepMonitor :→ i++,
    Method → {DifferentialEvolution, SearchPoints → sp, ,
    CrossProbability → cpScalingFactor → sc, PostProcess → False}];
  Clear[objfunc2];
  {Total[Flatten[mat1 ∗ permuteMatrix[mat2,
    Ordering[bestvec]]]], Ordering[bestvec]}]
```

We now show a run with printout included. The parameter settings are, as usual, based on shorter tuning runs.

Timing[QAP4[mat1, mat2, .08, 400, 320, .4, 4, .4, False, True]]

locally improved{4838, 4788}
new low 1 {iteration, elapsed time, new value} {0, 0.280017, 4788}
locally improved{4788, 4724}
new low 2 {iteration, elapsed time, new value} {0, 0.012001, 4724}
locally improved{4724, 4696}
new low 3 {iteration, elapsed time, new value} {0, 0., 4696}
locally improved{4696, 4644}
new low 4 {iteration, elapsed time, new value} {0, 0., 4644}
locally improved{4644, 4612}
new low 5 {iteration, elapsed time, new value} {0, 0.240015, 4612}
locally improved{4612, 4594}
new low 6 {iteration, elapsed time, new value} {0, 0.100006, 4594}
locally improved{4594, 4566}
new low 7 {iteration, elapsed time, new value} {0, 0.004, 4566}
locally improved{4566, 4498}
new low 8 {iteration, elapsed time, new value} {0, 0., 4498}
locally improved{4498, 4370}
new low 9 {iteration, elapsed time, new value} {0, 0.972061, 4370}
locally improved{4370, 4348}
new low 10 {iteration, elapsed time, new value} {0, 0.004, 4348}
locally improved{4348, 4322}
new low 11 {iteration, elapsed time, new value} {10, 21.3933, 4322}

new low 12 {iteration, elapsed time, new value} {11, 0.96806, 4308}
new low 13 {iteration, elapsed time, new value} {20, 19.0252, 4304}
locally improved{4304, 4242}
new low 14 {iteration, elapsed time, new value} {20, 1.88812, 4242}
locally improved{4242, 4184}
new low 15 {iteration, elapsed time, new value} {22, 4.29227, 4184}
new low 16 {iteration, elapsed time, new value} {29, 14.7769, 4174}
new low 17 {iteration, elapsed time, new value} {31, 4.57229, 4102}
locally improved{4102, 4096}
new low 18 {iteration, elapsed time, new value} {37, 12.3448, 4096}
locally improved{4096, 4092}
new low 19 {iteration, elapsed time, new value} {41, 8.0405, 4092}
new low 20 {iteration, elapsed time, new value} {51, 22.2414, 4082}
new low 21 {iteration, elapsed time, new value} {55, 8.28452, 4076}
new low 22 {iteration, elapsed time, new value} {56, 3.51622, 4072}
new low 23 {iteration, elapsed time, new value} {56, 0.396025, 3980}
new low 24 {iteration, elapsed time, new value} {62, 13.1488, 3964}
new low 25 {iteration, elapsed time, new value} {64, 3.03619, 3952}
locally improved{3952, 3948}
new low 26 {iteration, elapsed time, new value} {71, 16.385, 3948}
new low 27 {iteration, elapsed time, new value} {75, 8.38452, 3940}
new low 28 {iteration, elapsed time, new value} {78, 6.30839, 3934}
new low 29 {iteration, elapsed time, new value} {85, 14.0169, 3930}
new low 30 {iteration, elapsed time, new value} {85, 0.980061, 3924}
new low 31 {iteration, elapsed time, new value} {86, 1.71611, 3922}
locally improved{3922, 3894}
new low 32 {iteration, elapsed time, new value} {89, 7.22845, 3894}
locally improved{3894, 3870}
new low 33 {iteration, elapsed time, new value} {109, 42.1226, 3870}
new low 34 {iteration, elapsed time, new value} {119, 22.5814, 3860}
new low 35 {iteration, elapsed time, new value} {134, 33.4381, 3856}
locally improved{3856, 3840}
new low 36 {iteration, elapsed time, new value} {142, 16.269, 3840}
new low 37 {iteration, elapsed time, new value} {146, 8.72855, 3830}
new low 38 {iteration, elapsed time, new value} {174, 57.7716, 3816}
new low 39 {iteration, elapsed time, new value} {196, 44.6508, 3800}
new low 40 {iteration, elapsed time, new value} {203, 13.5768, 3788}
new low 41 {iteration, elapsed time, new value} {203, 0.400025, 3768}
locally improved{3768, 3750}
new low 42 {iteration, elapsed time, new value} {222, 34.3741, 3750}

{590.045, {3750,
 {1, 19, 22, 15, 13, 7, 10, 9, 20, 23, 21, 6, 14, 4, 17, 16, 3, 8, 25, 12, 24, 18, 11, 2, 5}}}

This is now quite close to the global minimum. As might be observed from the printout, the swaps occasionally let us escape from seemingly sticky local minima. So, for the problem at hand, this hybridization truly appears to confer an advantage over pure Differential Evolution. I will remark that it seems a bit more difficult to get this type of hybridization to cooperate well with the various shuffle methods of creating permutations.

For contrast we go to the opposite extreme and do a huge number of swaps, on a relatively smaller number of chromosomes and using far fewer iterations. We will reset our vector with swapped pairs to the original (or best variant found thereof, if we get improvements). This is to avoid straying far from reasonable vectors, since we now do many swaps.

This is thus far a 2-opt approach rather than Differential Evolution per se. Nonetheless, we notice that the later stages of improvement do come during the actual iterations of Differential Evolution, and quite possibly those final improvements are due in part to the maintaining of diversity and the use of mutation and recombination.

Timing[QAP4[mat1, mat2, .08, 20, 60, .4, 2000, .6, True, True]]

locally improved{4838, 4808}
new low 1 {iteration, elapsed time, new value} {0, 0.048003, 4808}
locally improved{4808, 4786}
new low 2 {iteration, elapsed time, new value} {0, 0.004001, 4786}
locally improved{4786, 4738}
new low 3 {iteration, elapsed time, new value} {0, 0., 4738}
locally improved{4738, 4690}
new low 4 {iteration, elapsed time, new value} {0, 0.004, 4690}
locally improved{4690, 4614}
new low 5 {iteration, elapsed time, new value} {0, 0., 4614}
locally improved{4614, 4502}
new low 6 {iteration, elapsed time, new value} {0, 0., 4502}
locally improved{4502, 4406}
new low 7 {iteration, elapsed time, new value} {0, 0., 4406}
locally improved{4406, 4370}
new low 8 {iteration, elapsed time, new value} {0, 0., 4370}
locally improved{4370, 4342}
new low 9 {iteration, elapsed time, new value} {0, 0.004, 4342}
locally improved{4342, 4226}
new low 10 {iteration, elapsed time, new value} {0, 0., 4226}
locally improved{4226, 4178}
new low 11 {iteration, elapsed time, new value} {0, 0.016001, 4178}
locally improved{4178, 4174}
new low 12 {iteration, elapsed time, new value} {0, 0., 4174}
locally improved{4174, 4170}
new low 13 {iteration, elapsed time, new value} {0, 0.016001, 4170}
locally improved{4170, 4158}
new low 14 {iteration, elapsed time, new value} {0, 0.012001, 4158}
locally improved{4158, 4114}
new low 15 {iteration, elapsed time, new value} {0, 0.004, 4114}

locally improved{4114,4070}
new low 16 {iteration, elapsed time, new value} {0,0.004,4070}
locally improved{4070,4046}
new low 17 {iteration, elapsed time, new value} {0,0.016001,4046}
locally improved{4046,4042}
new low 18 {iteration, elapsed time, new value} {0,0.060004,4042}
locally improved{4042,4014}
new low 19 {iteration, elapsed time, new value} {0,0.080005,4014}
locally improved{4014,3982}
new low 20 {iteration, elapsed time, new value} {0,0.008001,3982}
locally improved{3982,3978}
new low 21 {iteration, elapsed time, new value} {0,0.008,3978}
locally improved{3978,3970}
new low 22 {iteration, elapsed time, new value} {0,0.052003,3970}
locally improved{3970,3966}
new low 23 {iteration, elapsed time, new value} {0,0.096006,3966}
locally improved{3966,3964}
new low 24 {iteration, elapsed time, new value} {0,0.012001,3964}
locally improved{3964,3960}
new low 25 {iteration, elapsed time, new value} {0,0.,3960}
locally improved{3960,3944}
new low 26 {iteration, elapsed time, new value} {0,0.032002,3944}
locally improved{3944,3926}
new low 27 {iteration, elapsed time, new value} {0,0.036002,3926}
locally improved{3926,3916}
new low 28 {iteration, elapsed time, new value} {0,0.004001,3916}
locally improved{3916,3896}
new low 29 {iteration, elapsed time, new value} {0,0.032002,3896}
locally improved{3896,3892}
new low 30 {iteration, elapsed time, new value} {0,0.112007,3892}
locally improved{3892,3888}
new low 31 {iteration, elapsed time, new value} {0,0.096006,3888}
locally improved{3888,3868}
new low 32 {iteration, elapsed time, new value} {0,0.104006,3868}
locally improved{3868,3864}
new low 33 {iteration, elapsed time, new value} {0,2.18414,3864}
locally improved{3864,3860}
new low 34 {iteration, elapsed time, new value} {0,0.116007,3860}
locally improved{3860,3852}
new low 35 {iteration, elapsed time, new value} {0,1.84411,3852}
locally improved{3852,3838}
new low 36 {iteration, elapsed time, new value} {0,0.028002,3838}
locally improved{3838,3834}
new low 37 {iteration, elapsed time, new value} {0,0.016001,3834}
locally improved{3834,3818}

new low 38 {iteration, elapsed time, new value} {0, 0.072004, 3818}
locally improved{3818, 3812}
new low 39 {iteration, elapsed time, new value} {0, 3.55622, 3812}
locally improved{3812, 3786}
new low 40 {iteration, elapsed time, new value} {0, 0.084006, 3786}
locally improved{3786, 3780}
new low 41 {iteration, elapsed time, new value} {0, 1.6361, 3780}
locally improved{3780, 3768}
new low 42 {iteration, elapsed time, new value} {0, 0.048003, 3768}
locally improved{3768, 3758}
new low 43 {iteration, elapsed time, new value} {0, 0.096006, 3758}
locally improved{3758, 3756}
new low 44 {iteration, elapsed time, new value} {7, 315.692, 3756}
locally improved{3756, 3754}
new low 45 {iteration, elapsed time, new value} {10, 108.415, 3754}
locally improved{3754, 3752}
new low 46 {iteration, elapsed time, new value} {10, 0.15601, 3752}
locally improved{3752, 3748}
new low 47 {iteration, elapsed time, new value} {16, 245.787, 3748}
locally improved{3748, 3744}
new low 48 {iteration, elapsed time, new value} {16, 0.076005, 3744}

{872.687, {3744,
{22, 15, 20, 11, 5, 1, 9, 8, 25, 2, 19, 6, 3, 16, 18, 10, 7, 14, 21, 24, 13, 23, 4, 12, 17}}}

Notice that this permutation is not identical to the one we presented at the outset, which in turn comes from benchmark suite results in the literature. Also note that we seem to get good results from swaps early on (indeed, we almost get a global minimizer prior to the main iterations). This raises the question of whether it might be useful to plug in a different sort of heuristic, say larger swaps, or perhaps use of local (continuous) quadratic programming. The interested reader may wish to explore such possibilities.

4.8 Future Directions

We have seen several examples of discrete optimization problems, and indicated ways in which one might approach them using Differential Evolution. Problems investigated include basic integer programming, set partitioning, set covering by subsets, and the common permutation optimization problem of quadratic assignment. The main issues have been to adapt Differential Evolution to enforce discrete or combinatorial structure, e.g. that we obtain integrality, partitions, or permutations from chromosome vectors.

There are many open questions and considerable room for development. Here are a few of them.

- Figure out better ways to attack quadratic assignment problems so that we are less likely to encounter difficulty in tuning parameter values, premature convergence, and so on.

- Make the Differential Evolution program *adaptive*, that is, allow algorithm parameters themselves to be modified during the course of a run. This might make results less sensitive to tuning of parameters such as `CrossProbability`.
- Alternatively, develop a better understanding of how to select algorithm parameters in a problem-specific manner. Our experience has been that settings for cross probability should usually be around .9 (which is quite high as compared to what is typical for continuous optimization). It would be useful to have a more refined understanding of this and other tuning issues.
- Figure out how to sensibly alter parameters over the course of the algorithm, not by evolution but rather by some other measure, say iteration count. For example, one might do well to start of with a fairly even crossover (near 0.5, that is), and have it either go up toward 1, or drop toward 0, as the algorithm progresses. Obviously it is not hard to code Differential Evolution to do this. What might be interesting research is to better understand when and how such progression of algorithm parameters could improve performance.
- Implement a two-level version of Differential Evolution, wherein several short runs are used to generate initial values for a longer run.
- Use Differential Evolution in a hybridized form, say, with intermediate steps of local improvement. This would involving modifying chromosomes "in plac", so that improvements are passed along to subsequent generations. We showed a very basic version of this but surely there must be improvements to be found.

We remark that some ideas related to item 2 above are explored in [5]. Issues of self-adaptive tuning of Differential Evolution are discussed in some detail in [1]. A nice exposition of early efforts along these lines, for genetic algorithms, appears in [3].

References

1. Brest, J., Greiner, S., Bošković, B., Mernik, M., Žumer, V.: Self-adapting control parameters in differential evolution: a comparative study on numerical benchmark problems. IEEE Trans. Evol. Comput. 10, 646–657 (2006)
2. Gisvold, K., Moe, J.: A method for nonlinear mixed-integer programming and its application to design problems. J. ENg. Ind. 94, 353–364 (1972)
3. Goldberg, D.: Genetic Algorithms in Search, Optimization and Machine Learning. Addison-Wesley Longman Publishing Co., Inc., Boston (1989)
4. Goldberg, D., Lingle, R.: Alleles, loci, and the traveling salesman problem. In: Proceedings of the 1st International Conference on Genetic Algorithms, pp. 154–159. Lawrence Erlbaum Associates, Inc., Mahwah (1985)
5. Jacob, C.: Illustrating Evolutionary Computation with Mathematica. Morgan Kaufmann Publishers Inc., San Francisco (2001)
6. Krasnogor, N., Smith, J.: A tutorial for competent memetic algorithms: Model, taxonomy and design issues. IEEE Trans. Evol. Comput. 9, 474–488 (2005)
7. Lichtblau, D.: Discrete optimization using Mathematica. In: Proceedings of the World Conference on Systemics, Cybernetics, and Informatics. International Institute of Informatics and Systemics, vol. 16, pp. 169–174 (2000)

8. Mitchell, G., O'Donoghue, D., Barnes, D., McCarville, M.: GeneRepair: a repair operator for genetic algorithms. In: Cantú-Paz, E., Foster, J.A., Deb, K., Davis, L., Roy, R., O'Reilly, U.-M., Beyer, H.-G., Kendall, G., Wilson, S.W., Harman, M., Wegener, J., Dasgupta, D., Potter, M.A., Schultz, A., Dowsland, K.A., Jonoska, N., Miller, J., Standish, R.K. (eds.) GECCO 2003. LNCS, vol. 2724. Springer, Heidelberg (2003)
9. Moscato, P.: On evolution, search, optimization, genetic algorithms, and martial arts. Concurrent Computation Program 826, California Institute of Technology (1989)
10. Price, K., Storn, R.: Differential evolution. Dr. Dobb's Journal 78, 18–24 (1997)
11. Price, K., Storn, R., Lampinen, J.: Differential Evolution: A Practical Approach to Global Optimization. Natural Computing Series. Springer, New York (2005)
12. Tate, D., Smith, A.: A genetic approach to the quadratic assignment problem. Computers and Operations Research 22(1), 73–83 (1995)
13. Wolfram Research, Inc., Champaign, Illinois, USA, Mathematica 6 (2007) (Cited September 25, 2008),
http://reference.wolfram.com/mathematica/ref/NMinimize.html

5
Smallest Position Value Approach

Fatih Tasgetiren[1], Angela Chen[2], Gunes Gencyilmaz[3], and Said Gattoufi[4]

[1] Department of Operations Management and Business Statistics, Sultan Qaboos University, Muscat, Sultanate of Oman
mfatih@squ.edu.om
[2] Department of Finance, Nanya Institute of Technology, Taiwan 320, R.O.C
achen@nanya.edu.tw
[3] Department of Management, Istanbul Kultur University, Istanbul, Turkey
g.gencyilmaz@iku.edu.tr
[4] Department of Operations Management and Business Statistics, Sultan Qaboos University, Muscat, Sultanate of Oman
gattoufi@squ.edu.om

Abstract. In a traveling salesman problem, if the set of nodes is divided into clusters for a single node from each cluster to be visited, then the problem is known as the generalized traveling salesman problem (GTSP). Such problem aims to find a tour with minimum cost passing through only a single node from each cluster. In attempt to show how a continuous optimization algorithm can be used to solve a discrete/combinatorial optimization problem, this chapter presents a standard continuous differential evolution algorithm along with a smallest position value (SPV) rule and a unique solution representation to solve the GTSP. The performance of the differential evolution algorithm is tested on a set of benchmark instances with symmetric distances ranging from 51 (11) to 442 (89) nodes (clusters) from the literature. Computational results are presented and compared to a random key genetic algorithm (RKGA) from the literature.

5.1 Introduction

The generalized traveling salesman problem (GTSP), one of several variations of the traveling salesman problem (TSP), has been originated from diverse real life or potential applications. The TSP finds a routing of a salesman who starts from an origin (i.e. a home location), visits a prescribed set of cities, and returns to the origin in such a way that the total distance is minimum and each city is travelled once. On the other hand, in the GTSP, a salesman when making a tour does not necessarily visit all nodes. But similar to the TSP, the salesman will try to find a minimum-cost tour and travel each city exactly once. Since the TSP in its generality represents a typical NP-Hard combinatorial optimization problem, the GTSP is also NP-hard. While many other combinatorial optimization problems can be reduced to the GTSP problem [11], applications of the GTSP spans over several areas of knowledge including computer science, engineering, electronics, mathematics, and operations research, etc. For example, publications can be found in postal routing [11], computer file processing [9], order picking in warehouses [17], process planning for rotational parts [3], and the routing of clients through welfare agencies [24].

Let us first define the GTSP, a complete graph $G = (V,E)$ is a weighted undirected whose edges are associated with non-negative costs. We denote the cost of an edge $e = (i,j)$ by c_{ij}. Then, the set of V nodes is divided into m sets or clusters such that $V = \{V_1,..,V_m\}$ with $V = \{V_1 \cup .. \cup V_m\}$ and $V_j \cap V_k = \phi$. The problem involves two related decisions- choosing a node from the subset and finding a minimum cost tour in the subgraph of G. In other words, the objective is to find a minimum tour length containing exactly single node from each cluster V_j.

The GTSP was first addressed in [9, 24, 28]. Applications of various exact algorithms can be found in Laporte et al. [12, 13], Laporte & Nobert [11], Fischetti et al. [7, 8], and others in [4, 18]. Laporte & Nobert [11], developed an exact algorithm for GTSP by formulating an integer programming and finding the shortest Hamiltonian cycle through some clusters of nodes. Noon and Bean [18], presented a Lagrangean relaxation algorithm. Fischetti et al. [8] dealt with the asymmetric version of the problem and developed a branch and cut algorithm to solve this problem. While exact algorithms are very important, they are unreliable with respect to their running time which can easily reach many hours or even days, depending on the problem sizes. Meanwhile several other researchers use transformations from GTSP to TSP since a large variety of exact and heuristic algorithms have been applied for the TSP [3],. Lien et. al. [15] first introduced transformation of a GTSP into a TSP, where the number of nodes of the transformed TSP was very large. Then Dimitrijevic and Saric [6] proposed another transformation to decrease the size of the corresponding TSP. However, many such transformations depend on whether or not the problem is symmetric; moreover, while the known transformations usually allow to produce optimal GTSP tours from the obtained optimal TSP tours, such transformations do not preserve suboptimal solutions. In addition, such conversions of near-optimal TSP tours may result in infeasible GTSP solutions.

Because of the multitude of inputs and the time needed to produce best results, the GTSP problems are harder and harder to solve. That is why, in such cases, applications of several worthy heuristic approaches to the GTSP are considered. The most used construction heuristic is the nearest-neighbor heuristic which, in its adaptation form, was presented in Noon [17]. Similar adaptations of the farthest-insertion, nearest-insertion, and cheapest-insertion heuristics are proposed in Fischetti et al. [8]. In addition, Renaud & Boctor [20] developed one of the most sophisticated heuristics, called GI^3 (Generalized Initilialization, Insertion, and Improvement), which is a gen-eralization of the I^3 heuristic in Renaud et al. [21]. GI^3 contains three phases: in the Initialization phase, the node close to the other clusters is chosen from each cluster and greedily built into a tour that passes through some, but not necessarily all, of the chosen nodes. Next in the Insertion phase, nodes from unvisited clusters are inserted between two consecutive clusters on the tour in the cheapest possible manner, allowing the visited node to change for the adjacent clusters; after each insertion, the heuristic performs a modification of the 3-opt improvement method. In the Improvement phase, modifications of 2-opt and 3-opt are used to improve the tour. Here the modifications, called G2-opt, G3-opt, and G-opt, allow the visited nodes from each cluster to change as the tour is being re-ordered by the 2-opt or 3-opt procedures.

Application of evolutionary algorithms specifically to the GTSP have been few in the literature until Snyder & Daskin [26] who proposed a random key genetic algorithm

(RKGA) to solve this problem. In their RKGA, a random key representation is used and solutions generated by the RKGA are improved by using two local search heuristics namely, 2-opt and "swap". In the search process, their "swap" procedure is considered as a speed-up method which basically removes a node j from a tour and inserts all possible nodes ks from the corresponding cluster in between an edge (u,v) in a tour (i.e., between the node u and the node v). Such insertion is based on a modified nearest-neighbor criterion. These two local search heuristics have been separately embedded in the *level-I improvement* and *level-II improvement* procedures.

For each individual in the population, they store the original (pre-improvement) cost and the final cost after improvements have been made. When a new individual is created, they compare its pre-improvement cost to the pre-improvement cost of the individual at position $p \times N$ in the previous (sorted) population, where $p \in [0,1]$ is a parameter of the algorithm (they use $p = 0.05$ in their implementation). These two improvement procedures in Snyder & Daskin [26] are implemented as follows:

1. If the new solution is worse than the pre-improvement cost of this individual, the *level-I improvement* is considered. That is, one 2-opt exchange and one "swap" procedure (assuming a profitable one can be found) are performed and the resulting individual are stored.
2. Otherwise, the *level-II improvement* is considered. So the 2-opts are executed until no profitable 2-opts can be found, then the "swap" procedures are carried out until no profitable swaps can be found. The procedure is repeated until no further improvements have been made in a given pass.

The RKGA focuses on designing the local search to spend more time on improving solutions that seem promising to the previous solutions than the others. Both *level-I* and *level-II* improvements consider a "first-improvement" strategy, which means implementing the first improvement of a move, rather than the best improvement of such move.

Thereafter, Tasgetiren et al. [30, 31, 32] presented a discrete particle swarm optimization (**DPSO**) algorithm, a genetic algorithm (**GA**) and a hybrid iterated greedy (**HIG**) algorithm, respectively. They hybridized the above methods with a local search, called variable neighborhood descend algorithm, to further improve the solution quality; at the same time, they applied some speed-up methods for greedy node insertions. Silberholz & Golden proposed another **GA** in [25], which is denoted as **mrOXGA**.

Section 2 introduces a brief summary of discrete differential evolution algorithm. Section 3 provides the details of solution representation. Insertion methods are summarized in Section 4. Section 5 gives the details of the local search improvement heuristics. The computational results on benchmark instances are discussed in Section 6. Finally, Section 7 summarizes the concluding remarks.

5.2 Differential Evolution Algorithm

Differential evolution (DE) is a latest evolutionary optimization methods proposed by Storn & Price [27]. Like other evolutionary-type algorithms, DE is a population-based and stochastic global optimizer. The DE algorithm starts with establishing the initial population. Each individual has an **m**-dimensional vector with parameter values determined randomly and uniformly between predefined search ranges. In a DE algorithm, candidate solutions are represented by chromosomes based on floating-point numbers. In the mutation process of a DE algorithm, the weighted difference between two randomly selected population members is added to a third member to generate a mutated solution. Then, a crossover operator follows to combine the mutated solution with the target solution so as to generate a trial solution. Thereafter, a selection operator is applied to compare the fitness function value of both competing solutions, namely, target and trial solutions to determine who can survive for the next generation. Since DE was first introduced to solve the Chebychev polynomial fitting problem by Storn & Price [25], [27], it has been successfully applied in a variety of applications that can be found in Corne et. al [5], Lampinen [10], Babu & Onwubolu [1]; and Price et al. [19].

Currently, there are several variants of DE algorithms. We follow the *DE/rand/1/bin* scheme of Storn & Price [27] with the inclusion of SPV rule in the algorithm. Pseudocode of the DE algorithm is given in Fig 5.1.

>Initialize parameters
>Initialize the target population individuals
>Find the tour of the target population individuals
>Evaluate the target population individuals
>Apply local search to the target population individuals (Optional)
>Do{
>>Obtain the mutant population individuals
>>Obtain the trial population individuals
>>Find the tour of trial population individuals
>>Evaluate the trial population individuals
>>Do selection between the target and trial population individuals
>>Apply local searchto the target population individuals (Optional)
>
>}While (Not Termination)

Fig. 5.1. DE Algorithm with Local Search

The basic elements of DE algorithm are summarized as follows:

Target individual: X_i^t denotes the ith individual in the population at generation t and is defined as $X_i^t = [x_{i1}^t, x_{i2}^t, ..., x_{in}^t]$, where x_{ij}^t is the parameter value of the ith individual with respect to the jth dimension ($j = 1, 2, ..., n$).

Mutant individual: V_i^t denotes the ith individual in the population at generation t and is defined as $V_i^t = [v_{i1}^t, v_{i2}^t, ..., v_{in}^t]$, where v_{ij}^t is the parameter value of the ith individual with respect to the jth dimension ($j = 1, 2, ..., n$).

Trial individual: U_i^t denotes the ith individual in the population at generation t and is defined as $U_i^t = [u_{i1}^t, u_{i2}^t, ..., u_{in}^t]$, where u_{ij}^t is the parameter value of the ith individual with respect to the jth dimension ($j = 1, 2, ..., n$).

Target population: X^t is the set of *NP* individuals in the target population at generation t, i.e., $X^t = [X_1^t, X_2^t, ..., X_{NP}^t]$.

Mutant population: V^t is the set of *NP* individuals in the mutant population at generation t, i.e., $V^t = [V_1^t, V_2^t, ..., V_{NP}^t]$.

Trial population: U^t is the set of *NP* individuals in the trial population at generation t, i.e., $U^t = [U_1^t, U_2^t, ..., U_{NP}^t]$.

Tour: a newly introduced variable π_i^t, denoted a tour of the GTSP solution implied by the individual X_i^t, is represented as $\pi_i^t = [\pi_{i1}^t, \pi_{i2}^t, ..., \pi_{in}^t]$, where π_{ij}^t is the assignment of node j of the individual i in the tour at generation t.

Mutant constant: $F \in (0,2)$ is a real number constant which affects the differential variation between two individuals.

Crossover constant: $CR \in (0,1)$ is a real number constant which affects the diversity of population for the next generation.

Fitness function: In a minimization problem, the objective function is $f_i(\pi_i^t \leftarrow X_i^t)$, where π_i^t denotes the corresponding tour of individual X_i^t.

5.2.1 Solution Representation

In this section, we present a solution representation which enables **DEs** to solve the **GTSP**. Bean [2] suggested an encoding for the **GA** to solve the **GTSP**, where each set V_j has a gene consisting of an integer part between $[1, |V_j|]$ and a fractional part between $[0,1]$. The integer part indicates which node from the cluster is included in the tour, and the nodes are sorted by their fractional part to indicate the order. Similarly, a continuous **DE** can be used to solve the **GTSP**. First, we say each parameter value represents a cluster for the **GTSP** and is restricted to each cluster size of the GTSP instances.

From the following example, consider a GTSP instance with $V = \{1,..,20\}$ and $V_1 = \{1,...,5\}$, $V_2 = \{6,,...,10\}$, $V_3 = \{11,...,15\}$ and $V_4 = \{16,...,20\}$. The parameter values (x_j) can be positive or negative, e.g. for dimension j equal to 4.23 and for dimension j 2 is -3.07, etc. This feature indicates the difference between the random key encoding and the one in this chapter. Table 5.1 shows the solution representation of the DE for the GTSP. Then the integer parts of these parameter values (x_j) are respectively decoded as node 4 (the fourth node in V_1), node 8 (the third node in V_2), node 11 (the first node in V_3), and node 18 (the third node in V_1).

Table 5.1. Solution Representation

j	1	2	3	4
x_j	4.23	-3.07	1.80	3.76
v_j	4	8	11	18
s_j	0.23	-0.07	0.80	0.76
π_j	8	4	18	11
$F(\pi)$	$d_{8,4}$	$d_{4,18}$	$d_{18,11}$	$d_{11,8}$

Since the values of parameter x_j can be positive or negative, to determine which node (v_j) should be taken, the absolute value of the parameter x_j needs to be considered. Then the random key values (s_j) are determined by simply subtracting the integer part of the parameter x_j from its current value considering the negative signs, i.e., $s_j = x_j - \text{int}(x_j)$. So for dimension 1, its parameter x_1 is equal to 4.23 and the random key value s_1 is 0.23 ($S_1 = 4.23 - 4$). Finally, with respect to the random key values (s_j), the smallest position value (SPV) rule of Tasgetiren et. al. [29] is applied to the random key vector to determine the tour π. As illustrated in Table 5.1, the objective function value implied by a solution x with m nodes is the total tour length, which is given by

$$F(\pi) = \sum_{j=1}^{m-1} d_{\pi_j \pi_{j+1}} + d_{\pi_m \pi_1} \qquad (5.1)$$

However, with this proposed representation scheme, a problem may rise such that when the DE update equations are applied, any parameter value might be outside of the initial search range, which is restricted to the size of each cluster. Let $x_{\min}[j]$ and $x_{\max}[j]$ represent the minimum and maximum value of each parameter value for dimension j. Then they stand for the minimum and maximum cluster sizes of each dimension j. Regarding the initial population, each parameter value for the set V_j is drawn uniformly from $[-V_j + 1, V_j + 1]$. Obviously, $x_{\max}[j]$ is restricted to $[V_j + 1]$, whereas $x_{\min}[j]$ is restricted to $-x_{\max}[j]$. During the reproduction of the DE, when any parameter value is outside of the cluster size, it is randomly re-assigned to the corresponding cluster size again.

5.2.2 An Example Instance of the GTSP

In this section, we summarize the solution representation by using a GTSP instance of 11EIL51 from TSPLIB Library [23] with $V = \{1,..,51\}$, where the clusters are $V_1 = \{19,40,41\}$, $V_2 = \{3,20,35,36\}$, $V_3 = \{24,43\}$, $V_4 = \{33,39\}$, $V_5 = \{11,12,27,32,46,47,51\}$, $V_6 = \{2,16,21,29,34,50\}$, $V_7 = \{8,22,26,28,31\}$, $V_8 = \{13,14,18,25\}$,

Table 5.2. Clusters for the Instance 11EIL51

Cluster	Node						
V_1	19	40	41				
V_2	3	20	35	36			
V_3	24	43					
V_4	33	39					
V_5	11	12	27	32	46	47	51
V_6	2	16	21	29	34	50	
V_7	8	22	26	28	31		
V_8	13	14	18	25			
V_9	4	15	17	37	42	44	45
V_{10}	1	6	7	23	48		
V_{11}	5	9	10	30	38	49	

Table 5.3. Clusters for the Instance 11EIL51

Cluster					Node						
j	1	2	3	4	5	6	7	8	9	10	11
x_j	3.45	-2.66	1.86	1.11	-3.99	-6.24	-2.81	4.52	6.23	-1.89	3.02
v_j	41	20	24	33	27	50	22	25	44	1	10
s_j	0.45	-0.66	0.86	0.11	-0.99	-0.24	-0.81	0.52	0.23	-0.89	0.02
π_j	27	1	22	20	50	10	33	44	41	25	24
$d_{\pi_j \pi_{j+1}}$	$d_{27,1}$	$d_{1,22}$	$d_{22,20}$	$d_{20,50}$	$d_{50,10}$	$d_{10,33}$	$d_{33,44}$	$d_{44,41}$	$d_{41,25}$	$d_{25,24}$	$d_{24,27}$
$F(\pi)$	8	7	15	21	17	12	17	20	21	14	22

$V_9 = \{4, 15, 17, 37, 42, 44, 45\}$, $V_{10} = \{1, 6, 7, 23, 48\}$, and $V_{11} = \{5, 9, 10, 30, 38, 49\}$. To make clearer, we show the 11EIL51 instance in Table 5.2 below:

In order to establish the GTSP solution, each parameter value for the dimension j is restricted to each cluster size such that $-4 < x_1 < 4$, $-5 < x_2 < 5$, $-3 < x_3 < 3$, $-3 < x_4 < 3$, $-8 < x_5 < 8$, $-7 < x_6 < 7$, $-6 < x_7 < 6$, $-5 < x_8 < 5$, $-8 < x_9 < 8$, $-6 < x_{10} < 6$ and $-7 < x_{11} < 7$. This provides the feasibility of the GTSP solution generated by the DE algorithm. Suppose that a DE solution is obtained by the traditional update equations and the parameter values $x'_j s$ of the individual are given as in Table 5.3.

Similar to what we have explained via Table 5.1 example, the integer parts of the individual parameter values (x_j) are respectively decoded as node 41 (the third node in V_1), node 20 (the second node in V_2), node 24 (the first node in V_3), node 33 (the first node in V_4), node 27 (the third node in V_5), node 50 (the sixth node in V_6), node 22 (the second node in V_7), node 25 (the fourth node in V_8), node 44 (the sixth node in V_9), node 1 (the first node in V_{10}) and node 10 (the third node in V_{11}). Unlike the case in the RKGA, where the random key is defined as another vector, the fractional part of the individual parameter values (x_j) can be directly obtained as a random key to obtain the tour. As shown in Table 5.3, while applying the SPV rule to the random key vector (s_j), the tour (π_j) can be obtained very easily. As well, the objective function value of the individual X is given by

$$F(\pi) = \sum_{j=1}^{10} d_{\pi_j \pi_{j+1}} + d_{\pi_{10} \pi_1} = d_{27,1} + d_{1,22} + d_{22,20} + d_{20,50} + d_{50,10} + d_{10,33} + d_{33,44}$$
$$+ d_{44,41} + d_{41,25} + d_{25,24} + d_{24,27}$$

$$F(\pi) = \sum_{j=1}^{10} d_{\pi_j \pi_{j+1}} + d_{\pi_{11} \pi_1} = 8 + 7 + 15 + 21 + 17 + 12 + 17 + 20 + 21 + 14 + 22 = 174$$

5.2.3 Complete Computational Procedure of DE

The complete computational procedure of the DE algorithm for the GTSP problem can be summarized as follows:

- **Step 1: Initialization**
  - Set $t = 0$, $NP = 100$
  - Generate NP individuals randomly as in Table 5.1, $\{X_i^0, i = 1, 2, ..., NP\}$ where $X_i^0 = \left[x_{i1}^0, x_{i2}^0, ..., x_{in}^0\right]$.
  - Apply the SPV rule to find the tour $\pi_i^0 = \left[\pi_{i1}^0, \pi_{i2}^0, ..., \pi_{in}^0\right]$ of individual X_i^0 for $i = 1, 2, ..., NP$.
  - Evaluate each individual i in the population using the objective function f_i^0 $\left(\pi_i^0 \leftarrow X_i^0\right)$ for $i = 1, 2, ..., NP$.
- **Step 2: Update generation counter**
  - $t = t + 1$
- **Step 3: Generate mutant population**
  - For each target individual, X_i^t, $i = 1, 2, ..., NP$, at generation t, a mutant individual, $V_i^t = \left[v_{i1}^t, v_{i2}^t, ..., v_{in}^t\right]$, is determined such that:

$$V_i^t = X_{a_i}^{t-1} + F\left(X_{b_i}^{t-1} - X_{c_i}^{t-1}\right) \quad (5.2)$$

  where a_i, b_i and c_i are three randomly chosen individuals from the population such that $(a_i \neq b_i \neq c_i)$.
- **Step 4: Generate trial population**
  - Following the mutation phase, the crossover (re-combination) operator is applied to obtain the trial population. For each mutant individual, $V_i^t = \left[v_{i1}^t, v_{i2}^t, ..., v_{in}^t\right]$, an integer random number between 1 and n, i.e., $D_i \in (1, 2, ..., n)$, is chosen, and a trial individual, $U_i^t = \left[u_{i1}^t, u_{i2}^t, ..., u_{in}^t\right]$ is generated such that:

$$u_{ij}^t = \begin{cases} v_{ij}^t, & \text{if } r_{ij}^t \leq CR \text{ or } j = D_i \\ x_{ij}^{t-1}, & \text{Otherwise} \end{cases} \quad (5.3)$$

  where the index D refers to a randomly chosen dimension ($j = 1, 2, ..., n$), which is used to ensure that at least one parameter of each trial individual U_i^t differs from its counterpart in the previous generation U_i^{t-1}, CR is a user-defined crossover constant in the range (0, 1), and r_{ij}^t is a uniform random number between 0 and 1. In other words, the trial individual is made up with some parameters of mutant individual, or at least one of the parameters randomly selected, and some other parameters of target individual.
- **Step 5: Find tour**
  - Apply the SPV rule to find the tour $\pi_i^t = \left[\pi_{i1}^t, \pi_{i2}^t, ..., \pi_{in}^t\right]$ for $i = 1, 2, ..., NP$.
- **Step 6: Evaluate trial population**
  - Evaluate the trial population using the objective function $f_i^t \left(\pi_i^t \leftarrow U_i^t\right)$ for $i = 1, 2, ..., NP$.
- **Step 7: Selection**
  - To decide whether or not the trial individual U_i^t should be a member of the target population for the next generation, it is compared to its counterpart target individual X_i^{t-1} at the previous generation. The selection is based on the survival of fitness among the trial population and target population such that:

$$X_i^t = \begin{cases} U_i^t, & \text{if } f\left(\pi_i^t \leftarrow U_i^t\right) \leq f\left(\pi_i^{t-1} \leftarrow X_i^{t-1}\right) \\ X_i^{t-1}, & \text{otherwise} \end{cases} \quad (5.4)$$

- **Step 8: Stopping criterion**
  - If the number of generations exceeds the maximum number of generations, or some other termination criterion, then stop; otherwise go to step 2.

5.3 Insertion Methods

In this section of the chapter, the insertion methods denoted as **LocalSearchSD()** are modified from the literature and facilitate the use of the local search. Insertion methods are based on the insertion of node π_k^R into $m+1$ possible positions of a partial or destructed tour π^D with m nodes and an objective function value of $F\left(\pi^D\right)$. Note that as an example, only a single node is considered to be removed from the current solution to establish π_k^R with a single node and re-inserted into the partial solution. Such insertion of node π_k^R into $m-1$ possible positions is actually proposed by Rosenkrantz et al. [22] for the TSP. Snyder & Daskin [26] adopted it for the GTSP. It is based on the removal and the insertion of node π_k^R in an edge $\left(\pi_u^D, \pi_v^D\right)$ of a partial tour. However, it avoids the insertion of node π_k^R on the first and the last position of any given partial tour. Suppose that node $\pi_k^R=27$ will be inserted in a partial tour in Table 5.4.

Table 5.4. Partial Solution to Be Inserted for the Instance 11EIL51

j	1	2	3	4	5	6	7	8	9	10
π_j^D	1	22	20	50	10	33	44	41	25	24
$d_{\pi_j \pi_{j+1}}$	$d_{1,22}$	$d_{22,20}$	$d_{20,50}$	$d_{50,10}$	$d_{10,33}$	$d_{33,44}$	$d_{44,41}$	$d_{41,25}$	$d_{25,24}$	$d_{24,1}$
	173	7	15	21	17	12	17	20	21	14	29

A   Insertion of node π_k^R in the first position of the partial tour π^D
  a $Remove = d_{\pi_m^D \pi_1^D}$
  b $Add = d_{\pi_k^R \pi_1^D} + d_{\pi_m^D \pi_k^R}$
  b $F(\pi) = F\left(\pi^D\right) + Add - Remove$, where $F(\pi)$ and $F\left(\pi^D\right)$ are fitness function values of the tour after insertion and the partial tour, respectively.

Example A:

$Remove = d_{\pi_m^D \pi_1^D}$
$Remove = d_{\pi_{10}^D \pi_1^D}$
$Remove = d_{24,1}$
$Add = d_{\pi_k^R \pi_1^D} + d_{\pi_m^D \pi_k^R}$
$Add = d_{\pi_1^R \pi_1^D} + d_{\pi_{10}^D \pi_1^R}$
$Add = d_{27,1} + d_{24,27}$
$F(\pi) = F\left(\pi^D\right) + Add - Remove$

Table 5.5. Insertion of node $\pi_k^R=27$ into the first position of partial solution for Case A

j	1	2	3	4	5	6	7	8	9	10	11
π_j^D	27	1	22	20	50	10	33	44	41	25	24
$d_{\pi_j \pi_{j+1}}$	$d_{27,1}$	$d_{1,22}$	$d_{22,20}$	$d_{20,50}$	$d_{50,10}$	$d_{10,33}$	$d_{33,44}$	$d_{44,41}$	$d_{41,25}$	$d_{25,27}$	$d_{24,27}$
	174	8	7	15	21	17	12	17	20	21	14	22

$F(\pi) = d_{1,22} + d_{22,20} + d_{20,50} + d_{50,10} + d_{10,33} + d_{33,44} + d_{44,41} + d_{41,25} + d_{25,24} + d_{24,1} + d_{27,1} + d_{24,27} - d_{24,1}$

$F(\pi) = d_{1,22} + d_{22,20} + d_{20,50} + d_{50,10} + d_{10,33} + d_{33,44} + d_{44,41} + d_{41,25} + d_{25,24} + d_{27,1} + d_{24,27}$

B   Insertion of node π_k^R in the first position of the partial tour π^D
- a $Remove = d_{\pi_m^D \pi_1^D}$
- b $Add = d_{\pi_m^D \pi_k^R} + d_{\pi_k^R \pi_1^D}$
- b $F(\pi) = F(\pi^D) + Add - Remove$, where $F(\pi)$ and $F(\pi^D)$ are fitness function values of the tour after insertion and the partial tour, respectively.

Example B:

$Remove = d_{\pi_m^D \pi_1^D}$
$Remove = d_{\pi_{10}^D \pi_1^D}$
$Remove = d_{24,1}$
$Add = d_{\pi_m^D \pi_k^R} + d_{\pi_k^R \pi_1^D}$
$Add = d_{\pi_{10}^D \pi_1^R} + d_{\pi_1^R \pi_1^D}$
$Add = d_{24,27} + d_{27,1}$
$F(\pi) = F(\pi^D) + Add - Remove$

$F(\pi) = d_{1,22} + d_{22,20} + d_{20,50} + d_{50,10} + d_{10,33} + d_{33,44} + d_{44,41} + d_{41,25} + d_{25,24} + d_{24,1} + d_{24,27} + d_{27,1} - d_{24,1}$

$F(\pi) = d_{1,22} + d_{22,20} + d_{20,50} + d_{50,10} + d_{10,33} + d_{33,44} + d_{44,41} + d_{41,25} + d_{25,24} + d_{24,27} + d_{27,1}$

C   Insertion of node π_k^R between an edge (π_u^D, π_v^D)
- a $Remove = d_{\pi_u^D \pi_v^D}$
- b $Add = d_{\pi_u^D \pi_k^R} + d_{\pi_k^R \pi_v^D}$
- b $F(\pi) = F(\pi^D) + Add - Remove$, where $F(\pi)$ and $F(\pi^D)$ are fitness function values of the tour after insertion and the partial tour, respectively.

Table 5.6. Insertion of node $\pi_k^R=27$ into the last position of partial solution for Case B

j	1	2	3	4	5	6	7	8	9	10	11
π_j^D	1	22	20	50	10	33	44	41	25	24	27
$d_{\pi_j \pi_{j+1}}$	$d_{1,22}$	$d_{22,50}$	$d_{20,50}$	$d_{50,10}$	$d_{10,33}$	$d_{33,44}$	$d_{44,41}$	$d_{41,25}$	$d_{25,24}$	$d_{24,27}$	$d_{27,1}$
174	7	15	21	17	12	17	20	21	14	22	8

Example C:

$u = 6$
$v = 7$
$Remove = d_{\pi_u^D \pi_v^D}$
$Remove = d_{\pi_6^D \pi_7^D}$
$Remove = d_{33,44}$
$Add = d_{\pi_u^D \pi_k^R} + d_{\pi_k^R \pi_v^D}$
$Add = d_{\pi_6^D \pi_1^R} + d_{\pi_1^R \pi_7^D}$
$Add = d_{33,27} + d_{27,44}$
$F(\pi) = F(\pi^D) + Add - Remove$

$F(\pi) = d_{1,22} + d_{22,20} + d_{20,50} + d_{50,10} + d_{10,33} + d_{33,44} + d_{44,41} + d_{41,25} + d_{25,24} + d_{24,1} + d_{33,27} + d_{27,44} - d_{33,44}$

$F(\pi) = d_{1,22} + d_{22,20} + d_{20,50} + d_{50,10} + d_{10,33} + d_{44,41} + d_{41,25} + d_{25,24} + d_{24,1} + d_{33,27} + d_{27,44}$

Table 5.7. Insertion of node π_k^R between an edge (π_u^D, π_v^D) for Case C

j	1	2	3	4	5	6	7	8	9	10	11
π_j^D	1	22	20	50	10	33	27	44	41	25	24
$d_{\pi_j \pi_{j+1}}$	$d_{1,22}$	$d_{22,50}$	$d_{20,50}$	$d_{50,10}$	$d_{10,33}$	$d_{33,27}$	$d_{27,44}$	$d_{44,41}$	$d_{41,25}$	$d_{25,24}$	$d_{24,1}$
230	7	15	21	17	12	41	33	20	21	14	29

Note that **Case B** can actually be managed by **Case C**, since the tour is cyclic. Note again that the above insertion approach is somewhat different than the one in Snyder & Daskin [26], where the cost of an insertion of node π_k^R in an edge (π_u^D, π_v^D).

5.3.1 Hybridization with Local Search

The hybridization of DE algorithm with local search heuristics is achieved by performing a local search phase on every trial individual generated. The SWAP procedure [26], denoted as **LocalSearchSD** in this chapter, and the **2-opt** heuristic [21] were

$$\begin{aligned}
&\text{procedure } \text{LS}(\pi)\\
&\quad h := 1\\
&\quad \text{while } (h \leq m) \text{ do}\\
&\quad\quad \pi^* := LocalSearchSD(\pi)\\
&\quad\quad \text{if } (f(\pi^*) \leq f(\pi)) \text{ then}\\
&\quad\quad\quad \pi := \pi^*\\
&\quad\quad\quad h := 1\\
&\quad\quad \text{else}\\
&\quad\quad\quad h := h+1\\
&\quad\quad \text{else}\\
&\quad \text{endwhile}\\
&\quad \text{return } \pi\\
&\text{end procedure}
\end{aligned}$$

Fig. 5.2. The Local Search Scheme

Procedure SWAP()
remove j from T
for each $k \in V_j$
$\quad c^k \leftarrow \min\{d_{uk} + d_{kv} - d_{uv} / (u,v) \text{ is an edge in } T\}$
$\quad k^* \leftarrow \arg\min_{k \in V_j}\{c_k\}$
insert k^* into T between (u,v)

Fig. 5.3. The SWAP Procedure

separately applied to each trial individual. The **2-opt** heuristic finds two edges of a tour that can be removed and two edges that can be inserted in order to generate a new tour with a lower cost. More specifically, in the **2-opt** heuristic, the neighborhood of a tour is obtained as the set of all tours that can be replaced by changing two nonadjacent edges in that tour. Note that the **2-opt** heuristic is employed with the first improvement strategy in this study. The pseudo code of the local search (**LS**) procedures is given in Fig 5.2.

As to the **LocalSearchSD** procedure, it is based on the **SWAP** procedure and is basically concerned with removing a node from a cluster and inserting a different node from that cluster into the tour. The insertion is conducted using a modified nearest-neighbour criterion, so that the new node may be inserted on the tour in a spot different. Each node in the cluster is inserted into all possible spots in the current solution and the best insertion is replaced with the current solution. The **SWAP** procedure of Snyder & Daskin [26] is outlined in Fig 5.3, whereas the proposed DE algorithm is given in Fig 5.4. Note that in **SWAP** procedure, the followings are given such that tour T; set V_j; node $j \in V_j$, $j \in T$; distances d_{uv} between each $u, v \in V$.

5.4 Computational Results

Fischetti et al. [8] developed a branch-and-cut algorithm to solve the symmetric GTSP. The benchmark set is derived by applying a partitioning method to standard TSP

Procedure DE_GTSP
Set CR, F, NP, TerCriterion
$X = (x_1^0, x_2^0, ..., x_{NP}^0)$
$f(\pi_i^0 \leftarrow x_i^0)$
$\quad i:=1,2,...,NP$
$\pi_i^0 \leftarrow x_i^0 = 2\_opt(\pi_i^0 \leftarrow x_i^0)$
$\quad i=1,2,...,NP$
$\pi_i^0 \leftarrow x_i^0 = LS(\pi_i^0 \leftarrow x_i^0)$
$\quad i=1,2,...,NP$
$\pi_g^0 \leftarrow x_i^0 = \arg\min\{f(\pi_i^0 \leftarrow x_i^0)\}$
$\quad i=1,2,...,NP$
$\pi_B := \pi_g^0 \leftarrow x_i^0$
$k := 1$
while $(Not\ TerCriterion)\ do$
$\quad v_{ij}^k := x_{ia}^k + F(x_{ib}^k + x_{ic}^k)$
$\quad\quad i:=1,2,...,NP, j=1,2,...,m$
$\quad u_{ij}^k = \begin{cases} v_{ij}^k\ if\ r_{ij}^k < CR\ or\ j = D_j \\ x_{ij}^{k-1}\ otherwise \end{cases}$
$\quad\quad i=1,2,...,NP, j=1,2,...,m$
$\quad f(\pi_i^k \leftarrow u_i^k)$
$\quad\quad i:=1,2,...,NP$
$\quad \pi_i^k \leftarrow u_i^k = 2\_opt(\pi_i^k \leftarrow u_i^k)$
$\quad\quad i=1,2,...,NP$
$\quad \pi_i^k \leftarrow u_i^k = LS(\pi_i^k \leftarrow u_i^k)$
$\quad\quad i=1,2,...,NP$
$\quad x_i^k = \begin{cases} u_i^k\ if\ f(\pi_i^k \leftarrow u_i^k) < f(\pi_i^{k-1} \leftarrow x_i^{k-1}) \\ x_i^{k-1}\ otherwise \end{cases}$
$\quad\quad i=1,2,...,NP$
$\quad \pi_g^k \leftarrow x_g^k = \arg\min\{f(\pi_i^k \leftarrow x_i^k), f(\pi_g^{k-1} \leftarrow x_g^{k-1})\}$
$\quad\quad i:=1,2,...,NP$
$\quad \pi_B = \arg\min\{f(\pi_B), f(\pi_g^k \leftarrow x_g^k)\}$
$\quad k := k+1$
endwhile
return π_B

Fig. 5.4. The DE Algorithm with Local Search Heuristics

instances from the TSPLIB library [23]. The benchmark set with optimal objective function values for each of the problems is obtained through a personal communication with Dr. Lawrence V. Snyder. The benchmark set contains between 51 (11) and 442 (89) nodes (clusters) with Euclidean distances and the optimal objective function value for each of the problems is available. The DE algorithm was coded in Visual C++ and run on an Intel Centrino Duo 1.83 GHz Laptop with 512MB memory.

We consider the RKGA by Snyder & Daskin [26] for comparison in this paper due to the similarity in solution representation. The population size is taken as 100. Cross-over and mutation probability are taken as 0.9 and 0.2, respectively. To be consistent with Snyder & Daskin [26], the algorithm is terminated when 100 generations have been carried out or when 10 consecutive generations have failed to improve the best-known

Table 5.8. Computational Results of DE and RKGA Implementations

	DE		RKGA	
	F_{avg}	Δ_{avg}	F_{avg}	Δ_{avg}
11EIL51	219.4	26.1	227.4	30.7
14ST70	473.8	49.9	450.8	42.7
16EIL76	358.8	71.7	352	68.4
16PR76	93586.2	44.1	85385.2	31.5
20KROA100	20663	112.8	20191	107.9
20KROB100	20764.2	101	18537.4	79.5
20KROC100	20597.2	115.6	17871.6	87.1
20KROD100	19730.2	108.8	18477	95.5
20KROE100	20409.2	114.3	19787.6	107.8
20RAT99	1049	111.1	1090	119.3
20RD100	7349.2	101.3	7353.4	101.5
21EIL101	530.8	113.2	526.4	111.4
21LIN105	16170.2	96.9	14559.4	77.3
22PR107	64129.8	129.9	57724.6	106.9
25PR124	91609.4	150.3	82713	126
26BIER127	146725.2	102.6	154703.2	113.6
28PR136	115003.4	170.2	112674.6	164.7
29PR144	112725.6	145.7	94969.2	107
30KROA150	34961.8	217.3	31199.2	183.2
30KROB150	35184.8	188.5	34685.2	184.4
31PR152	140603.6	172.6	118813.4	130.4
32U159	61456.6	171.2	59099.2	160.8
39RAT195	3332	290.2	2844.2	233
40D198	30688.6	190.7	26453	150.6
40KROA200	49109.6	266.3	46866.4	249.6
40KROB200	48553.2	270.3	47303.2	260.8
45TS225	237888.4	248.1	229495.2	235.8
46PR226	259453.2	305.4	263699	312
53GIL262	4497	343.9	4233.6	314.8
53PR264	165646.6	460.6	145789.4	393.4
60PR299	116716.2	416.1	110977.8	390.2
64LIN318	98943.8	376.5	94469.2	352.1
80RD400	37058.6	482.6	34502.2	436.1
84FL417	68102	605.6	65025.6	573.5
88PR439	365437.8	508.1	364282.4	504.5
89PCB442	132388	511.3	131711.8	498

solution. Five runs were carried out for each problem instance to report the statistics based on the relative percent deviations (Δ) from optimal solutions as follows:

$$\Delta_{avg} = \sum_{i=1}^{R} \left(\frac{(H_i - OPT) \times 100}{OPT} \right) / R \qquad (5.5)$$

Table 5.9. Comparison for Optimal Instances of DE and RKGA Implementations

	DE		RKGA	
Instance	F_{avg}	Δ_{avg}	F_{avg}	Δ_{avg}
11EIL51	0	0.08	0	0.2
14ST70	0	0.1	0	0.2
16EIL76	0	0.12	0	0.2
16PR76	0	0.14	0	0.4
20KROA100	0	0.21	0	0.4
20KROB100	0	0.22	0	0.3
20KROC100	0	0.2	0	0.4
20KROD100	0	0.21	0	0.6
20KROE100	0	0.2	0	0.5
20RAT99	0	0.2	0	0.5
20RD100	0	0.2	0	0.4
21EIL101	0	0.19	0	0.5
21LIN105	0	0.21	0	0.4
22PR107	0	0.23	0	0.8
25PR124	0	0.28	0	0.4
26BIER127	0	0.33	0	0.5
28PR136	0	1.27	0	1
29PR144	0	0.37	0	0.7
30KROA150	0	0.48	0	0.9
30KROB150	0	0.46	0	1.2
31PR152	0.01	1.49	0	0.8
32U159	0	0.55	0	1
39RAT195	0.07	4.6	0	1.6
40D198	0.04	3.54	0	1.8
40KROA200	0	1.81	0	1.9
40KROB200	0.04	2.03	0	2.1
45TS225	0.25	2.98	0.02	1.5
46PR226	0	0.76	0	1.9
53GIL262	1.24	5.65	0.75	2.1
53PR264	0.01	4.38	0	3.2
60PR299	0.71	10.4	0.11	3.5
64LIN318	0.77	8.89	0.62	5.9
80RD400	1.64	18.89	1.19	5.3
84FL417	0.09	25.26	0.05	9.5
88PR439	1.13	22.94	0.27	9
89PCB442	1.78	12.12	1.7	1.72
Avg	0.22	3.67	0.13	0.2

where H_i, *OPT* and R are the objective function values generated by the DE in each run, the optimal objective function value, and the number of runs, respectively. For the computational effort consideration, t_{avg} denotes average CPU time in seconds to reach the best solution found so far during the run, i.e., the point of time that the best so far solution does not improve thereafter. F_{avg} represents the average objective function values out of five runs.

Table 5.8 shows the computational results of implementing DE without the local search methods and those adopted from Snyder & Daskin [26]. As seen in Table 5.8, the DE results are very competitive to the RKGA of Snyder & Daskin [26], even though a two-sided paired t-test favors the RKGA. However, our objective is just to show how a continuous optimization algorithm can be used for solving a combinatorial optimization problem. We would like to point out that with some better parameter tuning, the DE results could be further improved. In addition, our observation reveals the fact that the performance of the DE algorithm is tremendously affected by the mutation equation [14]. After applying the mutation operator, most dimension values fall outside of search limits (cluster sizes). To force them to be in the search range, they are randomly re-initialized between the search bounds in order to keep the DE algorithm search for nodes from clusters predefined. However, the random re-initialization causes the DE algorithm to conduct a random search, which ruins its learning ability. Based on our observation, using some different levels of crossover and mutation probabilities as well as other mutation operators did not have so much positive effect in the solution quality.

In spite of all the disadvantages above, the inclusion of local search improvement heuristics in Snyder & Daskin [26] has led the DE algorithm to be somehow competitive to the RKGA. The computational results with the local search heuristics are presented in Table 5.9.

As seen in Table 5.9, the DE algorithm with the local search improvement heuristics was able to generate competitive results to the RKGA of Snyder & Daskin [26]. However, as seen in both Table 5.8 and 5.9, the success was mainly due to the use of the local search improvement heuristics. A two-sided paired t-test on the relative percent deviations in Table 5.9 confirms that both DE and RKGA were statistically equivalent, since the p-value was 0.014. However, DE was computationally more expensive than RKGA.

5.5 Conclusions

A continuous DE algorithm is presented to solve the GTSP on a set of benchmark instances ranging from 51 (11) to 442 (89) nodes (clusters). The main contribution of this chapter is due to use of a continuous DE algorithm to solve a combinatorial optimization problem. For this reason, a unique solution representation is presented and the SPV rule is used to determine the tour. The pure DE algorithm without local search heuristics is competitive to RKGA. However, inclusion of the local search heuristics led the DE algorithm to be very competitive to the RKGA of Snyder & Daskin [26].

As we mentioned before, with some better parameter tuning, the DE results could have been further improved. However, our observation reveals the fact that the performance of the DE algorithm is tremendously affected by the mutation equation [14]. After applying the mutation operator, most parameter values fall outside of search limits (cluster sizes). To force them to be in the search range, they are randomly re-initialized between the search bounds in order to keep the DE algorithm search for nodes from clusters predefined. However, the random re-initialization causes the DE algorithm to conduct a random search, which ruins its learning ability. Based on our observation, using some different levels of crossover and mutation probabilities as well as other mutation operators did not have so much positive impact on the solution quality. In spite of all the disadvantages above, this work clearly shows the applicability of a continuous algorithm to solve a combinatorial optimization problem. .

For the future work, the current DE algorithm can be extended to solve some other combinatorial/discrete optimization problems based on clusters such as resource constrained project scheduling (mode selection), generalized assignment problem (agent selection), and so on. It will be also interesting to use the same representation for the particle swarm optimization and harmony search algorithms to solve the GTSP.

References

1. Babu, B., Onwubolu, G.: New Optimization Techniques in Engineering. Springer, Germany (2004)
2. Bean, J.: Genetic algorithms and random keys for sequencing and optimization. ORSA, Journal on Computing 6, 154–160 (1994)
3. Ben-Arieh, D., Gutin, G., Penn, M., Yeo, A., Zverovitch, A.: Process planning for rotational parts using the generalized traveling salesman problem. Int. J. Prod. Res. 41(11), 2581–2596 (2003)
4. Chentsov, A., Korotayeva, L.: The dynamic programming method in the generalized traveling salesman problem. Math. Comput. Model. 25(1), 93–105 (1997)
5. Corne, D., Dorigo, M., Glover, F.: Differential Evolution, Part Two. In: New Ideas in Optimization, pp. 77–158. McGraw-Hill, New York (1999)
6. Dimitrijevic, V., Saric, Z.: Efficient Transformation of the Generalized Traveling Salesman Problem into the Traveling Salesman Problem on Digraphs. Information Science 102, 65–110 (1997)
7. Fischetti, M., Salazar-Gonzalez, J., Toth, P.: The symmetrical generalized traveling salesman polytope. Networks 26(2), 113–123 (1995)
8. Fischetti, M., Salazar-Gonzalez, J., Toth, P.: A branch-and-cut algorithm for the symmetric generalized traveling salesman problem. Oper. Res. 45(3), 378–394 (1997)
9. Henry-Labordere, A.: The record balancing problem–A dynamic programming solution of a generalized traveling salesman problem. Revue Francaise D Informatique DeRecherche Operationnelle 3(NB2), 43–49 (1969)
10. Lampinen, J.: A Bibliography of Differential Evolution Algorithm, Technical Report, Lappeenranta University of Technology, Department of Information Technology. Laboratory of Information Processing (2000)
11. Laporte, G., Nobert, Y.: Generalized traveling salesman problem through n-sets of nodes - An integer programming approach. INFOR 21(1), 61–75 (1983)
12. Laporte, G., Mercure, H., Nobert, Y.: Finding the shortest Hamiltonian circuit through n clusters: A Lagrangian approach. Congressus Numerantium 48, 277–290 (1985)
13. Laporte, G., Mercure, H., Nobert, Y.: Generalized traveling salesman problem through n - sets of nodes - The asymmetrical case. Discrete Appl. Math. 18(2), 185–197 (1987)
14. Laporte, G., Asef-Vaziri, A., Sriskandarajah, C.: Some applications of the generalized traveling salesman problem. J. Oper. Res. Soc. 47(12), 461–1467 (1996)
15. Lien, Y., Ma, E., Wah, B.: Transformation of the Generalized Traveling Salesman Problem into the Standard Traveling Salesman Problem. Information Science 64, 177–189 (1993)
16. Lin, S., Kernighan, B.: An effective heuristic algorithm for the traveling salesman problem. Oper. Res. 21, 498–516 (1973)

17. Noon, C.: The generalized traveling salesman problem, Ph.D. thesis. University of Michigan (1988)
18. Noon, C., Bean, J.: A Lagrangian based approach for the asymmetric generalized traveling salesman problem. Oper. Res. 39(4), 623–632 (1991)
19. Price, K., Storn, R., Lapinen, J.: Differential Evolution - A Practical Approach to Global Optimization. Springer, Heidelberg (2006)
20. Renaud, J., Boctor, F.: An efficient composite heuristic for the symmetric generalized traveling salesman problem. Eur. J. OPer. Res. 108(3), 571–584 (1998)
21. Renaud, J., Boctor, F., Laporte, G.: A fast composite heuristic for the symmetric traveling salesman problem. INFORMS Journal on Computing 4, 134–143 (1996)
22. Rosenkrantz, D., Stearns, R., Lewis, P.: Approximate algorithms for the traveling salesman problem. In: Proceedings of the 15th annual symposium of switching and automata theory, pp. 33–42 (1974)
23. Reinelt, G.: TSPLIB. A travelling salesman problem library. ORSA Journal on Computing 4, 134–143 (1996)
24. Saskena, J.: Mathematical model of scheduling clients through welfare agencies. Journal of the Canadian Operational Research Society 8, 185–200 (1970)
25. Silberholz, J., Golden, B.: The generalized traveling salesman problem: A new genetic algorithm approach. In: Edward, K.B., et al. (eds.) Extending the horizons: Advances in Computing, Optimization and Decision Technologies, vol. 37, pp. 165–181. Springer, Heidelberg (1997)
26. Snyder, L., Daskin, M.: A random-key genetic algorithm for the generalized traveling salesman problem. Eur. J. Oper. Res. 174, 38–53 (2006)
27. Storn, R., Price, K.: Differential evolution - a simple and efficient adaptive scheme for global optimization over continuous spaces. ICSI, Technical Report TR-95-012 (1995)
28. Srivastava, S., Kumar, S., Garg, R., Sen, R.: Generalized traveling salesman problem through n sets of nodes. Journal of the Canadian Operational Research Society 7, 97–101 (1970)
29. Tasgetiren, M., Sevkli, M., Liang, Y.-C., Gencyilmaz, G.: Particle Swarm Optimization Algorithm for the Single Machine Total Weighted Tardiness Problem, In: The Proceeding of the World Congress on Evolutionary Computation, CEC 2004, pp. 1412–1419 (2004)
30. Tasgetiren, M., Suganthan, P., Pan, Q.-K.: A discrete particle swarm optimization algorithm for the generalized traveling salesman problem. In: The Proceedings of the 9th annual conference on genetic and evolutionary computation (GECCO 2007), London UK, pp. 158–167 (2007)
31. Tasgetiren, M., Suganthan, P., Pan, Q.-K., Liang, Y.-C.: A genetic algorithm for the generalized traveling salesman problem. In: The Proceeding of the World Congress on Evolutionary Computation (CEC 2007), Singapore, pp. 2382–2389 (2007)
32. Tasgetiren, M., Pan, Q.-K., Suganthan, P., Chen, A.: A hybrid iterated greedy algorithm for the generalized traveling salesman problem. Computers and Industrial Engineering (submitted, 2008)

6
Discrete/Binary Approach

Fatih Tasgetiren[1], Yun-Chia Liang[2], Quan-Ke Pan[3], and Ponnuthurai Suganthan[4]

[1] Department of Operations Management and Business Statistics, Sultan Qaboos University, Muscat, Sultanate of Oman
mfatih@squ.edu.om
[2] Department of Industrial Engineering and Management, Yuan Ze University, Taiwan
ycliang@saturn.yzu.edu.tw
[3] College of Computer Science, Liaocheng University, Liaocheng, P.R. China
qkpan@lctu.edu.cn
[4] School of Electrical and Electronic Engineering, Nanyang Technological University, Singapore 639798
epnsugan@ntu.edu.sg

Abstract. In a traveling salesman problem, if the set of nodes is divided into clusters so that a single node from each cluster can be visited, then the problem is known as the generalized traveling salesman problem where the objective is to find a tour with minimum cost passing through only a single node from each cluster. In this chapter, a discrete differential evolution algorithm is presented to solve the problem on a set of benchmark instances. The discrete differential evolution algorithm is hybridized with local search improvement heuristics to further improve the solution quality. Some speed-up methods presented by the authors previously are employed to accelerate the greedy node insertion into a tour. The performance of the hybrid discrete differential evolution algorithm is tested on a set of benchmark instances with symmetric distances ranging from 51 (11) to 1084 (217) nodes (clusters) from the literature. Computational results show its highly competitive performance in comparison to the best performing algorithms from the literature.

6.1 Introduction

The generalized traveling salesman problem (GTSP), one of several variations of the traveling salesman problem (TSP), has been originated from diverse real life or potential applications. The TSP finds a routing of a salesman who starts from an origin (i.e. a home location), visits a prescribed set of cities, and returns to the origin in such a way that the total distance is minimum and each city is travelled once. On the other hand, in the GTSP, a salesman when making a tour does not necessarily visit all nodes. But similar to the TSP, the salesman will try to find a minimum-cost tour and travel each city exactly once. Since the TSP in its generality represents a typical NP-Hard combinatorial optimization problem, the GTSP is also NP-hard. While many other combinatorial optimization problems can be reduced to the GTSP problem [11], applications of the GTSP spans over several areas of knowledge including computer science, engineering, electronics, mathematics, and operations research, etc. For example, publications can be found in postal routing [11], computer file processing [9], order picking in warehouses [18], process planning for rotational parts [3], and the routing of clients through welfare agencies [28].

Let us first define the GTSP, a complete graph $G = (V, E)$ is a weighted undirected whose edges are associated with non-negative costs. We denote the cost of an edge $e = (i, j)$ by d_{ij}. Then, the set of N nodes is divided into m sets or clusters such that $N = \{n_1, .., n_m\}$ with $N = \{n_1 \cup .. \cup n_m\}$ and $n_j \cap n_k = \phi$. The problem involves two related decisions- choosing a node from the subset and finding a minimum cost tour in the subgraph of G. In other words, the objective is to find a minimum tour length containing exactly single node from each cluster n_j.

The GTSP was first addressed in [9, 28, 32]. Applications of various exact algorithms can be found in Laporte et al. [12, 13], Laporte & Nobert [11], Fischetti et al. [7, 8], and others in [4, 19]. Laporte & Nobert [11], developed an exact algorithm for GTSP by formulating an integer programming and finding the shortest Hamiltonian cycle through some clusters of nodes. Noon and Bean [19], presented a Lagrangean relaxation algorithm. Fischetti et al. [8] dealt with the asymmetric version of the problem and developed a branch and cut algorithm to solve this problem. While exact algorithms are very important, they are unreliable with respect to their running time which can easily reach many hours or even days, depending on the problem sizes. Meanwhile several other researchers use transformations from GTSP to TSP since a large variety of exact and heuristic algorithms have been applied for the TSP [3],. Lien et. al. [15] first introduced transformation of a GTSP into a TSP, where the number of nodes of the transformed TSP was very large. Then Dimitrijevic and Saric [6] proposed another transformation to decrease the size of the corresponding TSP. However, many such transformations depend on whether or not the problem is symmetric; moreover, while the known transformations usually allow to produce optimal GTSP tours from the obtained optimal TSP tours, such transformations do not preserve suboptimal solutions. In addition, such conversions of near-optimal TSP tours may result in infeasible GTSP solutions.

Because of the multitude of inputs and the time needed to produce best results, the GTSP problems are harder and harder to solve. That is why, in such cases, applications of several worthy heuristic approaches to the GTSP are considered. The most used construction heuristic is the nearest-neighbor heuristic which, in its adaptation form, was presented in Noon [18]. Similar adaptations of the farthest-insertion, nearest-insertion, and cheapest-insertion heuristics are proposed in Fischetti et al. [8]. In addition, Renaud & Boctor [24] developed one of the most sophisticated heuristics, called GI^3 (Generalized Initilialization, Insertion, and Improvement), which is a gen-eralization of the I^3 heuristic in Renaud et al. [25]. GI^3 contains three phases: in the Initialization phase, the node close to the other clusters is chosen from each cluster and greedily built into a tour that passes through some, but not necessarily all, of the chosen nodes. Next in the Insertion phase, nodes from unvisited clusters are inserted between two consecutive clusters on the tour in the cheapest possible manner, allowing the visited node to change for the adjacent clusters; after each insertion, the heuristic performs a modification of the 3-opt improvement method. In the Improvement phase, modifications of 2-opt and 3-opt are used to improve the tour. Here the modifications, called G2-opt, G3-opt, and G-opt, allow the visited nodes from each cluster to change as the tour is being re-ordered by the 2-opt or 3-opt procedures.

Application of evolutionary algorithms specifically to the GTSP have been few in the literature until Snyder & Daskin [30] who proposed a random key genetic algorithm (RKGA) to solve this problem. In their RKGA, a random key representation is used and solutions generated by the RKGA are improved by using two local search heuristics namely, 2-opt and "swap". In the search process, their "swap" procedure is considered as a speed-up method which basically removes a node j from a tour and inserts all possible nodes ks from the corresponding cluster in between an edge (u,v) in a tour (i.e., between the node u and the node v). Such insertion is based on a modified nearest-neighbor criterion. These two local search heuristics have been separately embedded in the *level-I improvement* and *level-II improvement* procedures.

For each individual in the population, they store the original (pre-improvement) cost and the final cost after improvements have been made. When a new individual is created, they compare its pre-improvement cost to the pre-improvement cost of the individual at position $p \times N$ in the previous (sorted) population, where $p \in [0,1]$ is a parameter of the algorithm (they use $p = 0.05$ in their implementation). These two improvement procedures in Snyder & Daskin [30] are implemented as follows:

1. If the new solution is worse than the pre-improvement cost of this individual, the *level-I improvement* is considered. That is, one 2-opt exchange and one "swap" procedure (assuming a profitable one can be found) are performed and the resulting individual are stored.
2. Otherwise, the *level-II improvement* is considered. So the 2-opts are executed until no profitable 2-opts can be found, then the "swap" procedures are carried out until no profitable swaps can be found. The procedure is repeated until no further improvements have been made in a given pass.

The RKGA focuses on designing the local search to spend more time on improving solutions that seem promising to the previous solutions than the others. Both *level-I* and *level-II* improvements consider a "first-improvement" strategy, which means implementing the first improvement of a move, rather than the best improvement of such move.

Thereafter, Tasgetiren et al. [34, 35, 36] presented a discrete particle swarm optimization (**DPSO**) algorithm, a genetic algorithm (**GA**) and a hybrid iterated greedy (**HIG**) algorithm, respectively, whereas Silberholz & Golden proposed another **GA** in [29], which is denoted as **mrOXGA**.

Section 2 introduces a brief summary of discrete differential evolution algorithm. Section 3 provides the details of local search improvement heuristics. The computational results on benchmark instances are discussed in Section 4. Finally, Section 5 summarizes the concluding remarks.

6.2 Discrete Differential Evolution Algorithm

Differential evolution (DE) is a latest evolutionary optimization methods proposed by Storn & Price [31]. Like other evolutionary-type algorithms, DE is a population-based and stochastic global optimizer. The DE algorithm starts with establishing the initial population. Each individual has an m-dimensional vector with parameter values determined randomly and uniformly between predefined search ranges. In a DE algorithm, candidate solutions are represented by chromosomes based on floating-point numbers. In the mutation process of a DE algorithm, the weighted difference between two randomly selected population members is added to a third member to generate a mutated solution. Then, a crossover operator follows to combine the mutated solution with the target solution so as to generate a trial solution. Thereafter, a selection operator is applied to compare the fitness function value of both competing solutions, namely, target and trial solutions to determine who can survive for the next generation. Since DE was first introduced to solve the Chebychev polynomial fitting problem by Storn & Price [31], it has been successfully applied in a variety of applications that can be found in Corne et. al [5], Lampinen [19], Babu & Onwubolu [1]; and Price et al. [22].

Currently, there are several variants of DE algorithms. We follow the *DE/rand/1/bin* scheme of Storn & Price [31] with the inclusion of SPV rule in the algorithm. Pseudocode of the DE algorithm is given in Fig 6.1.

> Initialize parameters
> Initialize the target population individuals
> Find the tour of the target population individuals
> Evaluate the target population individuals
> Apply local search to the target population individuals (Optional)
> Do{
>   Obtain the mutant population individuals
>   Obtain the trial population individuals
>   Find the tour of trial population individuals
>   Evaluate the trial population individuals
>   Do selection between the target and trial population individuals
>   Apply local search to the target population individuals (Optional)
> }While (Not Termination)

Fig. 6.1. DE Algorithm with Local Search

The basic elements of DE algorithm are summarized as follows:

Target individual: X_i^k denotes the ith individual in the population at generation t and is defined as $X_i^k = [x_{i1}^k, x_{i2}^k, ..., x_{in}^k]$, where x_{ij}^k is the parameter value of the ith individual with respect to the jth dimension ($j = 1, 2, ..., m$).

Mutant individual: V_i^k denotes the ith individual in the population at generation t and is defined as $V_i^k = [v_{i1}^k, v_{i2}^k, ..., v_{in}^k]$, where v_{ij}^k is the parameter value of the ith individual with respect to the jth dimension ($j = 1, 2, ..., m$).

Trial individual: U_i^k denotes the ith individual in the population at generation t and is defined as $U_i^k = [u_{i1}^k, u_{i2}^k, ..., u_{in}^k]$, where u_{ij}^k is the parameter value of the ith individual with respect to the jth dimension ($j = 1, 2, ..., m$).

Target population: X^k is the set of *NP* individuals in the target population at generation t, i.e., $X^k = [X_1^k, X_2^k, ..., X_{NP}^k]$.

Mutant population: V^k is the set of *NP* individuals in the mutant population at generation *t*, i.e., $V^k = [V_1^k, V_2^k, ..., V_{NP}^k]$.

Trial population: U^k is the set of *NP* individuals in the trial population at generation *t*, i.e., $U^k = [U_1^k, U_2^k, ..., U_{NP}^k]$.

Mutant constant: $F \in (0,2)$ is a real number constant which affects the differential variation between two individuals.

Crossover constant: $CR \in (0,1)$ is a real number constant which affects the diversity of population for the next generation.

Fitness function: In a minimization problem, the objective function is given by $f_i(X_i^k)$, for the individual X_i^k.

Traditional DEs explained above are designed for continuous optimization problems where chromosomes are floating-point numbers. To cope with discrete spaces, a simple and novel discrete DE (DDE) algorithm is presented in [36, 20], where solutions are based on discrete/binary values. In the DDE algorithm, each target individual belonging to the *NP* number of individuals is represented by a solution as $X_i^k = [x_{i1}^k, x_{i2}^k, ..., x_{im}^k]$, consisting of discrete values of a permutation of clusters as well as a tour of nodes visited, at the generation *k*. The mutant individual is obtained by perturbing the generation best solution in the target population. So the differential variation is achieved in the form of perturbations of the best solution from the generation best solution in the target population. Perturbations are stochastically managed such that each individual in the mutant population is expected to be distinctive. To obtain the mutant individual, the following equation can be used:

$$V_i^k = \begin{cases} DC_d\left(X_g^{k-1}\right) & if\ r < P_m \\ insert\left(X_g^{k-1}\right) & otherwise \end{cases} \quad (6.1)$$

Where X_g^{k-1} is the best solution in the target population at the previous generation; P_m is the perturbation probability; DC_d is the destruction and construction procedure with the destruction size of *d* as a perturbation operator; and insert is a simple random insertion move from a given node to another node in the same cluster. A uniform random number *r* is generated between [0, 1]. If *r* is less than then the DC_d operator is applied to generate the mutant individual $V_i^k = DC_d\left(X_g^{k-1}\right)$; otherwise, the best solution from the previous generation is perturbed with a random insertion move resulting in the mutant individual $V_i^k = insert\left(X_g^{k-1}\right)$. Equation 6.1 will be denoted as $V_i^k := P_m \oplus DC_d\left(X_g^{k-1}\right)$ to ease the
$i:=1,2,...,NP$
understanding of pseudocodes. Following the perturbation phase, the trial individual is obtained such that:

$$U_i^k = \begin{cases} CR\left(V_i^k, X_i^{k-1}\right) & if\ r < P_c \\ V_i^k & otherwise \end{cases} \quad (6.2)$$

where *CR* is the crossover operator; and P_c is the crossover probability. In other words, if a uniform random number *r* is less than the crossover probability P_c, then the crossover operator is applied to generate the trial individual $U_i^k = CR\left(V_i^k, X_i^{k-1}\right)$. Otherwise the

trial individual is chosen as $U_i^k = V_i^k$. By doing so, the trial individual is made up either from the outcome of perturbation operator or from the crossover operator. Equation 6.2 will be denoted as $U_i^k := P_c \oplus CR\left(V_i^k, X_i^{k-1}\right)$.
$$i := 1,2,..,NP$$

Finally, the selection operator is carried out based on the survival of the fitness among the trial and target individuals such that:

$$X_i^k = \begin{cases} U_i^k & \text{if } f\left(\pi_i^k \leftarrow U_i^k\right) < f\left(\pi_i^{k-1} \leftarrow X_i^{k-1}\right) \\ X_i^{k-1} & \text{otherwise} \end{cases} \tag{6.3}$$

Equation 6.3 will be denoted as $X_i^k = \arg\min_{i:=1,2,...,NP} \left\{ f\left(\pi_i^k \leftarrow U_i^k\right), f\left(\pi_i^{k-1} \leftarrow X_i^{k-1}\right) \right\}$.

6.2.1 Solution Representation

We employ a path representation for the GTSP in this chapter. In the path representation, each consecutive node is listed in order. A disadvantage of this representation is due to the fact that there is no guarantee that a randomly selected solution will be a valid GTSP tour because there is no guarantee that each cluster is represented exactly once in the path without some repair procedures. To handle the GTSP, we include both cluster and tour information in the solution representation. The solution representation is illustrated in Table 6.1 where $d_{\pi_j \pi_{j+1}}$ shows the distance from node π_j to node π_{j+1}. Population individuals can be constructed in such a way that first a permutation of clusters is determined randomly, and then since each cluster contains one or more nodes, a tour is established by randomly choosing a single node from each corresponding cluster. For example, n_j stands for the cluster in the jth dimension, whereas π_j represents the node to be visited from the cluster n_j.

Table 6.1. Solution Representation

	j	1	...	m−1	m
	n_j	n_1	...	n_{m-1}	n_m
	π_j	π_1	...	π_{m-1}	π_m
X	$d_{\pi_j \pi_{j+1}}$	$d_{\pi_1 \pi_2}$	...	$d_{\pi_{m-1} \pi_m}$	$d_{\pi_m \pi_1}$
	$\sum_{j=1}^{m} d_{\pi_j \pi_{j+1}} + d_{\pi_m \pi_1}$	$d_{\pi_1 \pi_2}$	...	$d_{\pi_{m-1} \pi_m}$	$d_{\pi_m \pi_1}$

As illustrated in Table 6.1, the objective function value implied by a solution X with m nodes is the total tour length, which is given by

$$F(\pi) = \sum_{j=1}^{m-1} d_{\pi_j \pi_{j+1}} + d_{\pi_m \pi_1} \tag{6.4}$$

Table 6.2. Solution Representation

	j	1	2	3	4	5
	n_j	3	1	5	2	4
X	π_j	14	5	22	8	16
	$d_{\pi_j \pi_{j+1}}$	$d_{14,5}$	$d_{5,22}$	$d_{22,8}$	$d_{8,16}$	$d_{16,14}$

Now, consider a GTSP instance with $N = \{1,..,25\}$ where the clusters are $n_1 = \{1,..,5\}$, $n_2 = \{6,..,10\}$, $n_3 = \{11,..,15\}$, $n_4 = \{16,..,20\}$ and $n_5 = \{21,..,25\}$. Table 6.2 illustrates the example solution in detail.

So, the fitness function of the individual is given by $F(\pi) = d_{14,5} + d_{5,22} + d_{22,8} + d_{8,16} + d_{16,14}$.

6.2.2 Complete Computational Procedure of DDE

The complete computational procedure of the DDE algorithm for the GTSP problem can be summarized as follows:

- **Step 1: Initialization**
  - Set $t = 0$, $NP = 100$
  - Generate NP individuals randomly as in Table 6.2, $\{X_i^0, i = 1, 2, ..., NP\}$ where $X_i^0 = [x_{i1}^0, x_{i2}^0, ..., x_{im}^0]$.
  - Evaluate each individual i in the population using the objective function $f_i^0 (\pi_i^o \leftarrow X_i^0)$ for $i = 1, 2, ..., NP$.
- **Step 2: Update generation counter**
  - $k = k + 1$
- **Step 3: Generate mutant population**
  - For each target individual, X_i^k, $i = 1, 2, ..., NP$, at generation k, a mutant individual, $V_i^k = [v_{i1}^k, v_{i2}^k, ..., v_{im}^k]$, is determined such that:

  $$V_i^k = X_{a_i}^{k-1} + F\left(X_{b_i}^{k-1} - X_{c_i}^{k-1}\right) \quad (6.5)$$

  where a_i, b_i and c_i are three randomly chosen individuals from the population such that $(a_i \neq b_i \neq c_i)$.
- **Step 4: Generate trial population**
  - Following the mutation phase, the crossover (re-combination) operator is applied to obtain the trial population. For each mutant individual, $V_i^k = [v_{i1}^k, v_{i2}^k, ..., v_{im}^k]$, an integer random number between 1 and n, i.e., $D_i \in (1, 2, ..., m)$, is chosen, and a trial individual, $U_i^k = [u_{i1}^k, u_{i2}^t, ..., u_{im}^k]$ is generated such that:

  $$u_{ij}^k = \begin{cases} v_{ij}^k, & \text{if } r_{ij}^k \leq CR \text{ or } j = D_i \\ x_{ij}^{k-1}, & \text{Otherwise} \end{cases} \quad (6.6)$$

  where the index D refers to a randomly chosen dimension ($j = 1, 2, ..., m$), which is used to ensure that at least one parameter of each trial individual U_i^k differs from its counterpart in the previous generation U_i^{k-1}, CR is a user-defined crossover constant in the range (0, 1), and r_{ij}^k is a uniform random number between 0 and 1. In other words, the trial individual is made up with some parameters of mutant individual, or at least one of the parameters randomly selected, and some other parameters of target individual.

- **Step 5: Evaluate trial population**
  – Evaluate the trial population using the objective function $f_i^k\left(\pi_i^k \leftarrow U_i^k\right)$ for $i = 1,2,...,NP$.
- **Step 6: Selection**
  – To decide whether or not the trial individual U_i^k should be a member of the target population for the next generation, it is compared to its counterpart target individual X_i^{k-1} at the previous generation. The selection is based on the survival of fitness among the trial population and target population such that:

$$X_i^k = \begin{cases} U_i^k, & \text{if } f\left(\pi_i^k \leftarrow U_i^k\right) \leq f\left(\pi_i^{k-1} \leftarrow X_i^{k-1}\right) \\ X_i^{t-1}, & \text{otherwise} \end{cases} \quad (6.7)$$

- **Step 7: Stopping criterion**
  – If the number of generations exceeds the maximum number of generations, or some other termination criterion, then stop; otherwise go to step 2

6.2.3 NEH Heuristic

Due to the availability of the insertion methods from the TSP literature, which are modified in this chapter, it is possible to apply the NEH heuristic of Nawaz et al. [17] to the GTSP. Without considering cluster information for simplicity, the NEH heuristic for the GTSP can be summarized as follows:

1. Determine an initial tour of nodes. Let this tour be π.
2. The first two nodes (that is, π_1 and π_2) are chosen and two possible partial tours of these two nodes are evaluated. Note that since a tour must be a Hamiltonian cycle, partial tours will be evaluated with the first node being the last node as well. As an example, partial tours, (π_1, π_2, π_1) and (π_2, π_1, π_2) are evaluated first.
3. Repeat the following steps until all nodes are inserted. In the kth step, node π_k at position k is taken and tentatively inserted into all the possible k positions of the partial tour that are already partially completed. Select of these k tentative partial tours the one that results in the minimum objective function value or a cost function suitably predefined.

To picture out how the NEH heuristic can be adopted for the GTSP, consider a solution with five nodes as $\pi = \{3,1,4,2,5\}$. Following example illustrates the implementation of the NEH heuristic for the GTSP:

1. Current solution is $\pi = \{3,1,4,2,5\}$
2. Evaluate the first two nodes as follows: $\{3,1,3\}$ and $\{1,3,1\}$. Assume that the first partial tour has a better objective function value than the second one. So the current partial tour will be $\{3,1\}$.

3. Insertions:
    a) Insert node 4 into three possible positions of the current partial tour as follows: $\{4,3,1,4\}$, $\{3,4,1,3\}$ and $\{3,1,4,3\}$. Assume that the best objective function value is with the partial tour $\{3,4,1,3\}$. So the current partial tour will be $\{3,4,1\}$.
    b) Next, insert node 2 into four possible positions of the current partial tour as follows: $\{2,3,4,1,2\}$, $\{3,2,4,1,3\}$, $\{3,4,2,1,3\}$ and $\{3,4,1,2,3\}$. Assume that the best objective function value is with the partial tour $\{3,2,4,1,3\}$. So the current partial tour will be $\{3,2,4,1\}$.
    c) Finally, insert node 5 into five possible positions of the current partial tour as follows: $\{5,3,2,4,1,5\}$, $\{3,5,2,4,1,3\}$, $\{3,2,5,4,1,3\}$, $\{3,2,4,5,1,3\}$ and $\{3,2,4,1,5,3\}$. Assume that the best objective function value is with the partial tour $\{3,2,4,5,1,3\}$. So the final complete tour will be $\pi = \{3,2,4,5,1\}$.

6.2.4 Insertion Methods

In this section of the chapter, the insertion methods are modified from the literature and facilitate the use of the local search. It is important to note that for simplicity, we do not include the cluster information in the following examples. However, whenever an insertion move is carried out, the corresponding cluster is also inserted in the solution. Insertion methods are based on the insertion of node π_k^R into $m+1$ possible positions of a partial or destructed tour π^D with m nodes and an objective function value of $F\left(\pi^D\right)$. Note that as an example, only a single node is considered to be removed from the current solution to establish π_k^R with a single node and re-inserted into the partial solution. Such insertion of node π_k^R into $m-1$ possible positions is actually proposed by Rosenkrantz et al. [26] for the TSP. Snyder & Daskin [30] adopted it for the GTSP. It is based on the removal and the insertion of node π_k^R in an edge $\left(\pi_u^D, \pi_v^D\right)$ of a partial tour. However, it avoids the insertion of node π_k^R on the first and the last position of any given partial tour. Suppose that node $\pi_k^R = 8$ will be inserted in a partial tour in Table 6.3.

Table 6.3. Current solution

j	1	2	3	4	j	1
n_j^D	3	1	5	4	n_j^R	2
π_j^D	14	5	22	16	π_j^R	8
$d_{\pi_j \pi_{j+1}}$	$d_{14,5}$	$d_{5,22}$	$d_{22,16}$	$d_{16,14}$		

A   Insertion of node π_k^R in the first position of the partial tour π^D
     a $Remove = d_{\pi_m^D \pi_1^D}$
     b $Add = d_{\pi_k^R \pi_1^D} + d_{\pi_m^D \pi_k^R}$
     b $F(\pi) = F\left(\pi^D\right) + Add - Remove$, where $F(\pi)$ and $F\left(\pi^D\right)$ are fitness function values of the tour after insertion and the partial tour, respectively.

Table 6.4. Insertion of node $\pi_k^R=8$ in the first slot

j	1	2	3	4	5
n_j	2	3	1	5	4
π_j	8	14	5	22	16
$d_{\pi_j \pi_{j+1}}$	$d_{8,14}$	$d_{14,5}$	$d_{5,22}$	$d_{22,16}$	$d_{16,8}$

$Remove = d_{\pi_4 \pi_1} = d_{16,14}$
$Add = d_{\pi_u \pi_k} + d_{\pi_k \pi_v} = d_{14,8} + d_{8,5}$
$F(\pi) = F(\pi^D) + Add - Remove$
$F(\pi) = d_{14,5} + d_{5,22} + d_{22,16} + d_{16,14} + d_{8,14} + d_{16,8} - d_{16,14}$
$F(\pi) = d_{14,5} + d_{5,22} + d_{22,16} + d_{8,14} + d_{16,8}$

B   Insertion of node, pair π_k^R in the last position of the partial tour π^D
     a $Remove = d_{\pi_m^D \pi_1^D}$
     b $Add = d_{\pi_m^D \pi_k^R} + d_{\pi_k^R \pi_1^D}$
     b $F(\pi) = F(\pi^D) + Add - Remove$, where $F(\pi)$ and $F(\pi^D)$ are fitness function values of the tour after insertion and the partial tour, respectively.

Table 6.5. Insertion of node $\pi_k^R=8$ in the last slot

j	1	2	3	4	5
n_j	3	1	5	4	2
π_j	14	5	22	16	8
$d_{\pi_j \pi_{j+1}}$	$d_{14,5}$	$d_{5,22}$	$d_{22,16}$	$d_{16,8}$	$d_{8,14}$

$Remove = d_{\pi_4 \pi_1} = d_{16,14}$
$Add = d_{\pi_4 \pi_k} + d_{\pi_k \pi_1} = d_{16,8} + d_{8,14}$
$F(\pi) = F(\pi^D) + Add - Remove$
$F(\pi) = d_{14,5} + d_{5,22} + d_{22,16} + d_{16,14} + d_{16,8} + d_{8,14} - d_{16,14}$
$F(\pi) = d_{14,5} + d_{5,22} + d_{22,16} + d_{16,8} + d_{8,14}$

C   Insertion of node π_k^R between the edge $\left(\pi_u^D, \pi_v^D\right)$
     a $Remove = d_{\pi_u^D \pi_v^D}$
     b $Add = d_{\pi_u^D \pi_k^R} + d_{\pi_k^R \pi_v^D}$
     b $F(\pi) = F(\pi^D) + Add - Remove$, where $F(\pi)$ and $F(\pi^D)$ are fitness function values of the complete and the partial solutions respectively.

$Remove = d_{\pi_u \pi_v} = d_{14,5}$
$Add = d_{\pi_u \pi_k} + d_{\pi_k \pi_v} = d_{14,8} + d_{8,5}$
$F(\pi) = F(\pi^D) + Add - Remove$

Table 6.6. Insertion of node $\pi_k^R=8$ in between the edge (π_u^D, π_v^D)

j	1	2	3	4	5
n_j	3	2	1	5	4
π_j	14	8	5	22	16
$d_{\pi_j \pi_{j+1}}$	$d_{14,8}$	$d_{8,5}$	$d_{5,22}$	$d_{22,16}$	$d_{16,14}$

$F(\pi) = d_{14,5} + d_{5,22} + d_{22,16} + d_{14,8} + d_{8,5} - d_{33,44} + d_{44,41} + d_{41,25} + d_{25,24} + d_{14,5}$

$F(\pi) = d_{5,22} + d_{22,16} + d_{16,14} + d_{14,8} + d_{8,5}$

Note that **Case B** can actually be managed by **Case C**, since the tour is cyclic. Note again that the above insertion approach is somewhat different than the one in Snyder & Daskin [30], where the cost of an insertion of node π_k^R in an edge (π_u^D, π_v^D) is evaluated by $C = d_{\pi_u^D \pi_k^R} + d_{\pi_k^R \pi_v^D} - d_{\pi_u^D \pi_v^D}$. Instead, we directly calculate the fitness function value of the complete tour after using the insertion methods above, i.e., well suited for the NEH insertion heuristic..

6.2.5 Destruction and Construction Procedure

We employ the destruction and construction (DC) procedure of the iterated greedy (IG) algorithm [27] in the DDE algorithm. In the destruction step, a given number d of nodes, randomly chosen and without repetition, are removed from the solution. This results in two partial solutions. The first one with the size d of nodes is called X^R and includes the removed nodes in the order where they are removed. The second one with the size $m - d$ of nodes is the original one without the removed nodes, which is called X^D. It should be pointed out that we consider each corresponding cluster when the destruction and construction procedures are carried out in order to keep the feasibility of the GTSP tour. Note that the perturbation scheme is embedded in the destruction phase where p nodes from X^R are randomly chosen without repetition and they are replaced by some other nodes from the corresponding clusters.

The construction phase requires a constructive heuristic procedure. We employ the NEH heuristic described in the previous section. In order to reinsert the set X^R into the destructed solution X^D in a greedy manner, the first node π_1^R in X^R is inserted into all possible $m - d + 1$ positions in the destructed solution X^D generating $m - d + 1$ partial solutions. Among these $m - d + 1$ partial solutions including node π_1^R, the best partial solution with the minimum tour length is chosen and kept for the next iteration. Then the second node π_2^R in X^R is considered and so on until X^R is empty or a final solution is obtained. Hence X^D is again of size m.

The DC procedure for the GTSP is illustrated through Table 6.7 and Table 6.12 using the GTSP instance in Table 6.2. Note that the destruction size is $d = 2$ and the perturbation strength is $p = 1$ in this example. Perturbation strength $p = 1$ indicates replacing (mutating) only a single node among two nodes with another one from the same cluster.

Table 6.7. Current Solution

j	1	2	3	4	5
n_j	3	1	5	2	4
π_j	14	5	22	8	16

Table 6.8. Destruction Phase

j	1	**2**	3	**4**	5
n_j	3	**1**	5	**2**	4
π_j	14	**5**	22	**8**	16

Table 6.9. Destruction Phase

j	1	2	3	j	1	2
n_j^D	3	5	4	n_j^R	1	2
π_j^D	14	22	16	π_j^R	5	8

Table 6.10. Destruction Phase-Mutation

j	1	2	3	j	1	2
n_j^D	3	5	4	n_j^R	1	**2**
π_j^D	14	22	16	π_j^R	5	**9**

Table 6.11. Construction Phase

j	1	2	3	4	j	1
n_j^D	3	5	1	4	n_j^R	2
π_j^D	14	22	5	16	π_j^R	9

Table 6.12. Final Solution

j	1	2	3	4	5
n_j^D	3	**2**	5	**1**	4
π_j^D	14	**9**	22	**5**	16

Table 6.13. Two-Cut PTL Crossover Operator

	j	1	2	3	4	5
P1	n_j	5	1	4	2	3
	π_j	24	3	19	8	14
P2	n_j	5	1	4	2	3
	π_j	24	3	19	8	14
O1	n_j	4	2	5	1	3
	π_j	19	8	24	3	14
O2	n_j	5	1	3	4	2
	π_j	24	3	14	19	8

Step 1.a. Choose $d = 2$ nodes with corresponding clusters, randomly.
Step 1.b. Establish $\pi^D = \{14, 22, 16\}$, $n^D = \{3, 5, 4\}$, $\pi^R = \{5, 8\}$ and $n^R = \{1, 2\}$.
Step 1.c. Perturb $\pi^R = \{5, 8\}$ to $\pi^R = \{5, 9\}$ by randomly choosing $n_2^R = 2$ in the set $n^R = \{1, 2\}$, and randomly replacing $n_2^R = 8$ with $n_2^R = 9$ from the same cluster n_2.
Step 2.a. After the best insertion of node $\pi_1^R = 5$ and the cluster $\pi_1^R = 1$.
Step 2.b. After the best insertion of node $n_2^R = 9$ and the cluster $\pi_1^R = 2$.
$F(\pi) = d_{5,22} + d_{22,16} + d_{16,14} + d_{14,8} + d_{8,5}$

6.2.6 PTL Crossover Operator

Two-cut PTL crossover operator developed by Pan et al. [21] is used in the DDE algorithm. The two-cut PTL crossover operator is able to produce a pair of distinct offspring even from two identical parents. An illustration of the two-cut PTL cross-over operator is shown in Table 6.13.

In the PTL crossover, a block of nodes and clusters from the first parent is determined by two cut points randomly. This block is either moved to the right or left corner of the offspring. Then the offspring is filled out with the remaining nodes and corresponding clusters from the second parent. This procedure will always produce two distinctive offspring even from the same two parents as shown in Table 6.13. In this chapter, one of these two unique offspring is chosen randomly with an equal probability of 0.5.

6.2.7 Insert Mutation Operator

Insert mutation operator is a modified insert mutation considering the clusters in the solution representation. It is also used in the perturbation of the solution in the destruction and construction procedure. It is basically related to removing a node from a tour of an individual, and replacing that particular node with another one from the same cluster. It is illustrated in Table 6.14.

As shown in Table 6.14, the cluster $n_2 = 5$ is randomly selected and its corresponding node $\pi_2 = 23$ is replaced by node $\pi_2 = 22$ from the same cluster $n_2 = 5$.

Table 6.14. Insert Mutation

	j	1	2	3	4	5
X_i	n_j	3	5	2	1	4
	π_j	12	23	8	4	19
X_i	n_j	3	5	2	1	4
	π_j	12	22	8	4	19

6.2.8 DDE Update Operations

To figure out how the individuals are updated in the DDE algorithm, an example using the GTSP instance in Table 6.2 is also illustrated through Table 6.15 and Table 6.18. Assume that the mutation and crossover probabilities are 1.0, the two-cut PTL crossover

Table 6.15. An Example of Individual Update

	j	1	2	3	4	5
X_i	n_j	3	5	2	1	4
	π_j	12	23	8	4	19
G_i	n_j	3	1	5	2	4
	π_j	15	4	24	7	17

Table 6.16. Insert Mutation

	j	1	2	3	4	5
X_i	n_j	3	1	5	2	4
	π_j	15	4	24	7	17
G_i	n_j	3	1	5	2	4
	π_j	15	4	25	7	17

Table 6.17. Two-Cut PTL Crossover

	j	1	2	3	4	5
X_i	n_j	3	5	2	1	4
	π_j	12	23	8	4	19
G_i	n_j	3	*1*	5	2	4
	π_j	15	*4*	25	7	17
U_i	n_j	2	1	3	5	4
	π_j	8	4	15	25	17

Table 6.18. Selection For Next Generation

		j	1	2	3	4	5
X_i	n_j		3	5	2	1	4
	π_j		12	23	8	4	19
U_i	n_j		2	1	3	5	4
	π_j		8	4	15	25	17
		Assume that $f(U_i) \leq f(X_i)$ $X_i = U_i$					
U_i	n_j		2	1	3	5	4
	π_j		8	4	15	25	17

and insert mutation operators are employed. Given the individual and the global best (best so far solution for DDE) solution, the global best solution is first mutated by using equation 6.1. For example, in Table 6.15, the dimensions $u=3$ is chosen randomly with its corresponding cluster and node. Node $\pi_u = \pi_3 = 25$ is replaced by $\pi_u = \pi_3 = 24$ from the same cluster $n_u = n_3 = 5$, thus resulting in the mutant individual V_i. Then the mutant individual V_i is recombined with its corresponding individual X_i in the target population to generate the trial individual U_i by using equation 6.2. Finally, the target individual X_i is compared to the trial individual U_i to determine which one would survive for the next generation based on the survival of the fittest by using equation 6.3.

6.3 Hybridization with Local Search

The hybridization of DE algorithm with local search heuristics is achieved by performing a local search phase on every trial individual generated. The SWAP procedure [30], denoted as **LocalSearchSD** in this chapter, and the **2-opt** heuristic [16] were separately applied to each trial individual. The **2-opt** heuristic finds two edges of a tour that can be removed and two edges that can be inserted in order to generate a new tour with a lower cost. More specifically, in the **2-opt** heuristic, the neighborhood of a tour is obtained as the set of all tours that can be replaced by changing two nonadjacent edges in that tour. Note that the **2-opt** heuristic is employed with the first improvement strategy in this study. The pseudo code of the local search (**LS**) procedures is given in Fig 6.2.

As to the **LocalSearchSD** procedure, it is based on the **SWAP** procedure and is basically concerned with removing a node from a cluster and inserting a different node from that cluster into the tour. The insertion is conducted using a modified nearest-neighbour criterion, so that the new node may be inserted on the tour in a spot different. Each node in the cluster is inserted into all possible spots in the current solution and the best insertion is replaced with the current solution. The **SWAP** procedure of Snyder & Daskin [30] is outlined in Fig 6.3, whereas the proposed DDE algorithm is given in Fig 6.4. Note that in **SWAP** procedure, the followings are given such that tour T; set V_j; node $j \in V_j$, $j \in T$; distances d_{uv} between each $u, v \in V$.

```
procedure  LS (π ← X)
    h := 1
    while  (h ≤ m) do
        π* := LocalSearchSD (π ← X)
        if  (f (π* ← X) ≤ f (π ← X))  then
            π ← X := π* ← X
            h := 1
        else
            h := h + 1
        else
    endwhile
    return  π ← X
end procedure
```

Fig. 6.2. The Local Search Scheme

```
Procedure  SWAP ()
remove  j  from  T
for each  k ∈ V_j
    c^k ← min {d_uk + d_kv − d_uv / (u,v) is an edge in T }
    k* ← arg min {c_k}
          k∈V_j
insert  k*  into T between (u, v)
```

Fig. 6.3. The SWAP Procedure

6.4 Computational Results

Fischetti et al. [8] developed a branch-and-cut algorithm to solve the symmetric GTSP. The benchmark set is derived by applying a partitioning method to standard TSP instances from the TSPLIB library [23]. The benchmark set with optimal objective function values for each of the problems is obtained through a personal communication with Dr. Lawrence V. Snyder. The benchmark set contains between 51 (11) and 442 (89) nodes (clusters) with Euclidean distances and the optimal objective function value for each of the problems is available. The DDE algorithm was coded in Visual C++ and run on an Intel Centrino Duo 1.83 GHz Laptop with 512MB memory. The population size was fixed at 100. The initial population is constructed randomly and then the NEH insertion heuristic was applied to each random solution. The destruction size and perturbation strength were taken as 5 and 3, respectively. The crossover and mutation probability were taken as 0.9 and 0.2, respectively. PTL [33] crossover operator is used in the DDE algorithm. The DDE algorithm was terminated when the best so far solution was not improved after 50 consecutive generations. Five runs were carried out for each problem instance to report the statistics based on the relative percent deviations(Δ) from optimal solutions as follows

$$\Delta_{avg} = \sum_{i=1}^{R} \left(\frac{(H_i - OPT) \times 100}{OPT} \right) / R \qquad (6.8)$$

Procedure DE_GTSP
Set $CR, F, NP, TerCriterion$
$X = \left(x_1^0, x_2^0, ..., x_{NP}^0\right)$
$f\left(\pi_i^0 \leftarrow x_i^0\right)$
$i := 1, 2, ..., NP$
$\pi_i^0 \leftarrow x_i^0 = 2\_opt\left(\pi_i^0 \leftarrow x_i^0\right)$
$i = 1, 2, ..., NP$
$\pi_i^0 \leftarrow x_i^0 = LS\left(\pi_i^0 \leftarrow x_i^0\right)$
$i = 1, 2, ..., NP$
$\pi_g^0 \leftarrow x_i^0 = \arg\min\left\{f\left(\pi_i^0 \leftarrow x_i^0\right)\right\}$
$i = 1, 2, ..., NP$
$\pi_B := \pi_g^0 \leftarrow x_i^0$
$k := 1$
while $(Not\ TerCriterion)$ *do*
$\quad V_{ij}^k := x_{ia}^k + F\left(X_{ib}^k + X_{ic}^k\right)$
$\quad i := 1, 2, ..., NP, j = 1, 2, ..., m$
$\quad u_{ij}^k = \begin{cases} v_{ij}^k\ if\ r_{ij}^k < CR\ or\ j = D_j \\ x_{ij}^{k-1}\ otherwise \end{cases}$
$\quad i = 1, 2, ..., NP, j = 1, 2, ..., m$
$\quad f\left(\pi_i^k \leftarrow U_i^k\right)$
$\quad i := 1, 2, ..., NP$
$\quad \pi_i^k \leftarrow U_i^k = 2\_opt\left(\pi_i^k \leftarrow U_i^k\right)$
$\quad i = 1, 2, ..., NP$
$\quad \pi_i^k \leftarrow U_i^k = LS\left(\pi_i^k \leftarrow U_i^k\right)$
$\quad i = 1, 2, ..., NP$
$\quad X_i^k = \begin{cases} u_i^k\ if\ f\left(\pi_i^k \leftarrow u_i^k\right) < f\left(\pi_i^{k-1} \leftarrow X_i^{k-1}\right) \\ x_i^{k-1}\ otherwise \end{cases}$
$\quad i = 1, 2, ..., NP$
$\quad \pi_g^k \leftarrow X_g^k = \arg\min\left\{f\left(\pi_i^k \leftarrow X_i^k\right), f\left(\pi_g^{k-1} \leftarrow X_g^{k-1}\right)\right\}$
$\quad i := 1, 2, ..., NP$
$\quad \pi_B \leftarrow X_B = \arg\min\left\{f\left(\pi_B \leftarrow X_B\right), f\left(\pi_g^k \leftarrow X_g^k\right)\right\}$
$\quad k := k + 1$
endwhile
return $\pi_B \leftarrow X_B$

Fig. 6.4. The DDE Algorithm with Local Search Heuristics

where H_i, *OPT* and R are the objective function values generated by the DDE in each run, the optimal objective function value, and the number of runs, respectively. For the computational effort consideration, t_{avg} denotes average CPU time in seconds to reach the best solution found so far during the run, i.e., the point of time that the best so far solution does not improve thereafter. F_{avg} represents the average objective function values out of five runs.

6.4.1 Solution Quality

Table 6.19 gives the computational results for each of the problem instances in detail. As seen in Table 6.19, the DDE algorithm was able to obtain optimal solutions in at least two of the five runs for 35 out of 36 problems tested (97%). For 32 (89%) out of

Table 6.19. Computational Results of DDE algorithm

Instance	OPT	n_{opt}	Δ_{avg}	Δ_{min}	Δ_{max}	I_{avg}	I_{min}	I_{max}	t_{avg}	t_{min}	t_{max}
11EIL51	174	5	0	0	0	1	1	1	0.04	0.02	0.06
14ST70	316	5	0	0	0	1	1	1	0.04	0.03	0.05
16EIL76	209	5	0	0	0	1	1	1	0.05	0.05	0.06
16PR76	64925	5	0	0	0	1	1	1	0.06	0.05	0.06
20KROA100	9711	5	0	0	0	1	1	1	0.09	0.08	0.09
20KROB100	10328	5	0	0	0	1	1	1	0.09	0.08	0.09
20KROC100	9554	5	0	0	0	1	1	1	0.08	0.08	0.09
20KROD100	9450	5	0	0	0	1	1	1	0.08	0.08	0.09
20KROE100	9523	5	0	0	0	1	1	1	0.09	0.08	0.09
20RAT99	497	5	0	0	0	1	1	1	0.08	0.08	0.09
20RD100	3650	5	0	0	0	1	1	1	0.09	0.08	0.09
21EIL101	249	5	0	0	0	1	1	1	0.08	0.08	0.09
21LIN105	8213	5	0	0	0	1	1	1	0.1	0.09	0.11
22PR107	27898	5	0	0	0	1	1	1	0.1	0.09	0.11
25PR124	36605	5	0	0	0	1	1	1	0.13	0.13	0.14
26BIER127	72418	5	0	0	0	1	1	1	0.14	0.13	0.14
28PR136	42570	5	0	0	0	1	1	1	0.18	0.16	0.19
29PR144	45886	5	0	0	0	1	1	1	0.18	0.17	0.2
30KROA150	11018	5	0	0	0	1	1	1	0.2	0.19	0.2
30KROB150	12196	5	0	0	0	1	1	1	0.2	0.19	0.2
31PR152	51576	5	0	0	0	1.2	1	2	0.22	0.19	0.28
32U159	22664	5	0	0	0	1	1	1	0.23	0.22	0.24
39RAT195	854	5	0	0	0	1.4	1	2	0.42	0.36	0.48
40D198	10557	5	0	0	0	1.4	1	2	0.44	0.38	0.52
40KROA200	13406	5	0	0	0	1.2	1	2	0.41	0.38	0.48
40KROB200	13111	5	0	0	0	7	1	22	0.93	0.41	2.03
45TS225	68340	3	0.04	0	0.09	9.8	1	33	1.32	0.47	3.05
46PR226	64007	5	0	0	0	1	1	1	0.42	0.41	0.44
53GIL262	1013	2	0.41	0	0.69	11.4	1	44	2	0.72	5.36
53PR264	29549	5	0	0	0	1.4	1	3	0.79	0.67	1.23
60PR299	22615	2	0.05	0	0.09	11.2	6	19	3.24	2.5	5.36
64LIN318	20765	5	0	0	0	14.2	3	45	4.37	2.13	10.28
80RD400	6361	5	0	0	0	14.8	11	18	8.3	6.86	9.97
84FL417	9651	3	0.01	0	0.02	13.8	8	24	6.86	4.58	10.88
88PR439	60099	5	0	0	0	15.2	8	23	8.54	6.06	11.08
89PCB442	21657	5	0	0	0	19	10	35	11.72	7.86	17.8
Overal Avg		4.72	0.01	0	0.02	4.03	2.11	8.22	1.45	1	2.27

36 problems, the DDE algorithm obtained the optimal solution in *every* trial. The DDE algorithm solved all the problems with a 0.01% deviation on average, 0.00% deviation on minimum and 0.02% deviation on maximum. The overall hit ratio was 4.72, which indicates that the DDE algorithm was able to find the 95% of the optimal solutions on overall average. The worst case performance was never more than 0.02% above optimal

on overall average. In other words, it indicates that the DDE algorithm has a tendency of yielding consistent solutions across a wide variety of problem instances. To highlight its consistency more, the range between the best and worst case was only 0.02% on overall average, thus indicating a very robust algorithm.

6.4.2 Computation Time

Table 6.19 also gives necessary information about CPU time requirement for each of the problem instances. The DDE algorithm is very fast in terms of CPU time requirement due to the mean CPU time of less than 12 seconds for all instances. In addition, its maximum CPU time was no more than 18 seconds for all instances. The DDE algorithm was able to find its best solution in the first 4.03 generations on overall average and spent most of the time waiting for the termination condition. It took 2.11 generations at minimum and only 8.22 generations at maximum on overall average to find its best solution for each problem instance. Since the local search heuristics are applied to each problem instance at each generation, most of the running times have been devoted to the local search improvement heuristics, which indicates the impact of the them on the solution quality. It implies the fact that with some better local search heuristics such as Renaud and Boctor's G2-opt or G3-opt, as well as with some speed-up methods for 2-opt heuristic, its CPU time performance may be further improved.

6.4.3 Comparison to Other Algorithms

We compare the DDE algorithm to two genetic algorithms, namely, RKGA by Snyder & Daskin [30] and mrOXGA by Silberholz & Golden [29], where RKGA is re-implemented under the same machine environment. Table 6.20 summarizes the solution quality in terms of relative percent deviations from the optimal values and CPU time requirements for all three algorithms. Note that our machine has a similar speed as Silberholz & Golden [29]. A two-sided paired t-test which compares the results on Table 6.20 with a null hypothesis that the algorithms were identical generated p-values of 0.175 and 0.016 for DDE vs. mrOXGA and DDE vs. RKGA,respectively, suggesting near-identical results between DDE and mrOXGA. On the other hand, the paired t-test confirms that the differences between DDE and RKGA were significant on the behalf of DDE subject to the fact that DDE was computationally less expensive than both RKGA and mrOXGA since p-values were 0.001 for DDE vs. mrOXGA and 0.008 for DDE vs. RKGA.

In addition to above, Silberholz & Golden [29] provided larger problem instances ranging from 493 (99) to 1084 (217) nodes (clusters) where no optimal solutions are available. However, they provided the results of mrOXGA and RKGA. We compare the DDE results to those presented in Silberholz & Golden [29]. As seen in Table 6.21, DDE generated consistently better results than both RKGA and mrOXGA in terms of both solution quality and CPU time requirement even if the larger instances are considered. In particular, 8 out 9 larger instances are further improved by the DDE algorithm. The paired t-test on the objective function values on Table 6.21 confirms that the differences between DDE and RKGA were significant since p-value was 0.033 (null hypothesis is rejected), whereas DDE was equivalent to mrOXGA since p-value was 0.237. In

Table 6.20. Comparison for Optimal Instances

Instance	DDE		mrOXGA		RKGA	
	Δ_{avg}	t_{avg}	Δ_{avg}	t_{avg}	Δ_{avg}	t_{avg}
11EIL51	0	0.04	0	0.26	0	0.08
14ST70	0	0.04	0	0.35	0	0.07
16EIL76	0	0.05	0	0.37	0	0.11
16PR76	0	0.06	0	0.45	0	0.16
20KROA100	0	0.09	0	0.63	0	0.25
20KROB100	0	0.09	0	0.6	0	0.22
20KROC100	0	0.08	0	0.62	0	0.23
20KROD100	0	0.08	0	0.67	0	0.43
20KROE100	0	0.09	0	0.58	0	0.15
20RAT99	0	0.08	0	0.5	0	0.24
20RD100	0	0.09	0	0.51	0	0.29
21EIL101	0	0.08	0	0.48	0	0.18
21LIN105	0	0.1	0	0.6	0	0.33
22PR107	0	0.1	0	0.53	0	0.2
25PR124	0	0.13	0	0.68	0	0.26
26BIER127	0	0.14	0	0.78	0	0.28
28PR136	0	0.18	0	0.79	0.16	0.36
29PR144	0	0.18	0	1	0	0.44
30KROA150	0	0.2	0	0.98	0	0.32
30KROB150	0	0.2	0	0.98	0	0.71
31PR152	0	0.22	0	0.97	0	0.38
32U159	0	0.23	0	0.98	0	0.55
39RAT195	0	0.42	0	1.37	0	1.33
40D198	0	0.44	0	1.63	0.07	1.47
40KROA200	0	0.41	0	1.66	0	0.95
40KROB200	0	0.93	0.05	1.63	0.01	1.29
45TS225	0.04	1.32	0.14	1.71	0.28	1.09
46PR226	0	0.42	0	1.54	0	1.09
53GIL262	0.41	2	0.45	3.64	0.55	3.05
53PR264	0	0.79	0	2.36	0.09	2.72
60PR299	0.05	3.24	0.05	4.59	0.16	4.08
64LIN318	0	4.37	0	8.08	0.54	5.39
80RD400	0	8.3	0.58	14.58	0.72	10.27
84FL417	0.01	6.86	0.04	8.15	0.06	6.18
88PR439	0	8.54	0	19.06	0.83	15.09
89PCB442	0	11.72	0.01	23.43	1.23	11.74
Avg	0.01	1.45	0.04	2.99	0.13	2

Table 6.21. Comparision to Silberholz & Golden-Time is milliseconds

	DE		mrOXGA		RKGA	
Instance	F_{avg}	t_{avg}	F_{avg}	t_{avg}	F_{avg}	t_{avg}
11EIL51	**174**	37.6	**174**	259.2	**174**	78.2
14ST70	**316**	43.8	**316**	353	**316**	65.6
16EIL76	**209**	50	**209**	369	**209**	106.4
16PR76	**64925**	56.4	**64925**	447	**64925**	156.2
20KROA100	**9711**	90.6	**9711**	628.2	**9711**	249.8
20KROB100	**10328**	87.6	**10328**	603.2	**10328**	215.6
20KROC100	**9554**	84.4	**9554**	621.8	**9554**	225
20KROD100	**9450**	81.2	**9450**	668.8	**9450**	434.4
20KROE100	**9523**	87.6	**9523**	575	**9523**	147
20RAT99	**497**	81.2	**497**	500	**497**	243.8
20RD100	**3650**	90.6	**3650**	506.2	**3650**	290.8
21EIL101	**249**	81.2	**249**	478.2	**249**	184.6
21LIN105	**8213**	96.8	**8213**	603.2	**8213**	334.4
22PR107	**27898**	96.8	27898.6	534.4	27898.6	197
25PR124	**36605**	134.2	**36605**	678	**36605**	259
26BIER127	**72418**	137.4	**72418**	784.4	**72418**	275.2
28PR136	**42570**	175	**42570**	793.8	42639.8	362.8
29PR144	**45886**	184.2	**45886**	1003.2	45887.4	437.6
30KROA150	**11018**	200	**11018**	981.2	**11018**	319
30KROB150	**12196**	200	**12196**	978.4	**12196**	712.4
31PR152	**51576**	218.8	**51576**	965.4	**51576**	381.2
32U159	**22664**	228.2	**22664**	984.4	**22664**	553.2
39RAT195	**854**	415.6	**854**	1374.8	**854**	1325
40D198	**10557**	437.6	**10557**	1628.2	10564	1468.6
40KROA200	**13406**	412.4	**13406**	1659.4	**13406**	950.2
40KROB200	**13111**	931.2	13117.6	1631.4	13112.2	1294.2
45TS225	**68364**	1322	68435.2	1706.2	68530.8	1087.4
46PR226	**64007**	421.8	**64007**	1540.6	**64007**	1094
53GIL262	**1017.2**	2000	1017.6	3637.4	1018.6	3046.8
53PR264	**29549**	793.8	**29549**	2359.4	29574.8	2718.6
60PR299	**22627**	3243.6	**22627**	4593.8	22650.2	4084.4
64LIN318	**20765**	4368.8	**20765**	8084.4	20877.8	5387.6
80RD400	**6361**	8303.2	6397.8	14578.2	6407	10265.6
84FL417	**9651.6**	6856.4	9654.6	8152.8	9657	6175.2
88PR439	**60099**	8543.6	**60099**	19059.6	60595.4	15087.6
89PCB442	**21657**	11718.8	21658.2	23434.4	21923	11743.8
99D493	**20059.2**	15574.8	20117.2	35718.8	20260.4	14887.8
115RAT575	**2421**	20240.2	2414.8	48481	2442.4	46834.4
131P654	**27430**	30428.4	27508.2	32672	27448.4	46996.8
132D657	**22544.8**	57900	22599	132243.6	22857.6	58449.8
145U724	**17367.2**	74687.4	17370.6	161815.2	17806.2	59625.2
157RAT783	**3272.2**	77000.2	3300.2	152147	3341	89362.4
201PR1002	114692.8	211025.2	**114582.2**	464356.4	117421.2	332406.2
212U1060	**106460**	247187.4	108390.4	594637.4	110158	216999.8
217VM1084	**131718.2**	292381.6	131884.6	562040.6	133743.4	390115.6
Overal Avg	**27502.7**	**23971.9**	27554.28	50930.41	27741.29	29503.03

terms of CPU times, the paired t-test on the CPU times confirms that the differences between DDE and mrOXGA were significant since the p-values was 0.020, whereas it was failed to reject the null hypothesis of being equal difference between DDE and RKGA due to the p-value of 0.129. Briefly, the paired t-test indicates that DDE was able to generate lower objective function values with less CPU times than mrOXGA. On the other hand, DDE yielded much better objective function values with identical CPU times than RKGA.

6.5 Conclusions

A DDE algorithm is presented to solve the GTSP on a set of benchmark instances ranging from 51 (11) to 1084 (217) nodes (clusters). The contributions of this paper can be summarized as follows. A unique solution representation including both cluster and tour information is presented, which handles the GTSP properly when carrying out the DDE operations. To the best of our knowledge, this is the first reported application of the DDE algorithm applied to the GTSP. The perturbation scheme is presented in the destruction procedure. Furthermore, the DDE algorithm is donated with very effective local search methods, 2-opt and SWAP procedure, in order to further improve the solution quality. Ultimately, the DDE algorithm was able to find optimal solutions for a large percentage of problem instances from a set of test problems from the literature. It was also able to further improve 8 out of 9 larger instances from the literature. Both solution quality and computation times are competitive to or even better than the best performing algorithms from the literature. In particular, its performance on the larger instances is noteworthy.

Acknowledgement

P. Suganthan acknowledges the financial support offered by the A*Star (Agency for Science, Technology and Research) under the grant # 052 101 0020.

References

1. Babu, B., Onwubolu, G.: New Optimization Techniques in Engineering. Springer, Germany (2004)
2. Bean, J.: Genetic algorithms and random keys for sequencing and optimization. ORSA, Journal on Computing 6, 154–160 (1994)
3. Ben-Arieh, D., Gutin, G., Penn, M., Yeo, A., Zverovitch, A.: Process planning for rotational parts using the generalized traveling salesman problem. Int. J. Prod. Res. 41(11), 2581–2596 (2003)
4. Chentsov, A., Korotayeva, L.: The dynamic programming method in the generalized traveling salesman problem. Math. Comput. Model. 25(1), 93–105 (1997)
5. Corne, D., Dorigo, M., Glover, F.: Differential Evolution, Part Two. In: New Ideas in Optimization, pp. 77–158. McGraw-Hill, New York (1999)
6. Dimitrijevic, V., Saric, Z.: Efficient Transformation of the Generalized Traveling Salesman Problem into the Traveling Salesman Problem on Digraphs. Information Science 102, 65–110 (1997)

7. Fischetti, M., Salazar-Gonzalez, J., Toth, P.: The symmetrical generalized traveling salesman polytope. Networks 26(2), 113–123 (1995)
8. Fischetti, M., Salazar-Gonzalez, J., Toth, P.: A branch-and-cut algorithm for the symmetric generalized traveling salesman problem. Oper. Res. 45(3), 378–394 (1997)
9. Henry-Labordere, A.: The record balancing problem–A dynamic programming solution of a generalized traveling salesman problem. Revue Francaise D Informatique DeRecherche Operationnelle 3(NB2), 43–49 (1969)
10. Lampinen, J.: A Bibliography of Differential Evolution Algorithm, Technical Report, Lappeenranta University of Technology, Department of Information Technology. Laboratory of Information Processing (2000)
11. Laporte, G., Nobert, Y.: Generalized traveling salesman problem through n-sets of nodes - An integer programming approach. INFOR 21(1), 61–75 (1983)
12. Laporte, G., Mercure, H., Nobert, Y.: Finding the shortest Hamiltonian circuit through n clusters: A Lagrangian approach. Congressus Numerantium 48, 277–290 (1985)
13. Laporte, G., Mercure, H., Nobert, Y.: Generalized traveling salesman problem through n - sets of nodes - The asymmetrical case. Discrete Appl. Math. 18(2), 185–197 (1987)
14. Laporte, G., Asef-Vaziri, A., Sriskandarajah, C.: Some applications of the generalized traveling salesman problem. J. Oper. Res. Soc. 47(12), 461–1467 (1996)
15. Lien, Y., Ma, E., Wah, B.: Transformation of the Generalized Traveling Salesman Problem into the Standard Traveling Salesman Problem. Information Science 64, 177–189 (1993)
16. Lin, S., Kernighan, B.: An effective heuristic algorithm for the traveling salesman problem. Oper. Res. 21, 498–516 (1973)
17. Nawaz, M., Enscore, E., Ham, I.: Heuristic algorithm for the m-machine, n-job flow shop sequencing problem. OMEGA 11(1), 91–95 (1983)
18. Noon, C.: The generalized traveling salesman problem, Ph.D. thesis. University of Michigan (1988)
19. Noon, C., Bean, J.: A Lagrangian based approach for the asymmetric generalized traveling salesman problem. Oper. Res. 39(4), 623–632 (1991)
20. Pan, Q.-K., Tasgetiren, M., Liang, Y.-C.: A Discrete differential evolution algorithm for the permutation flowshop scheduling problem. Comput. Ind. Eng. (2008)
21. Pan, Q.-K., Tasgetiren, M., Liang, Y.-C.: A Discrete Particle Swarm Optimization Algorithm for the No-Wait Flowshop Scheduling Problem with Makespan and Total Flowtime Criteria. Comput. Oper. Res. 35, 2807–2839 (2008)
22. Price, K., Storn, R., Lapinen, J.: Differential Evolution - A Practical Approach to Global Optimization. Springer, Heidelberg (2006)
23. Reinelt, G.: TSPLIB. A travelling salesman problem library. ORSA Journal on Computing 4, 134–143 (1996)
24. Renaud, J., Boctor, F.: An efficient composite heuristic for the symmetric generalized traveling salesman problem. Eur. J. Oper. Res. 108(3), 571–584 (1998)
25. Renaud, J., Boctor, F., Laporte, G.: A fast composite heuristic for the symmetric traveling salesman problem. INFORMS Journal on Computing 4, 134–143 (1996)
26. Rosenkrantz, D., Stearns, R., Lewis, P.: Approximate algorithms for the traveling salesman problem. In: Proceedings of the 15th annual symposium of switching and automata theory, pp. 33–42 (1974)
27. Ruiz, R., Stützle, T.: A simple and effective iterated greedy algorithm for the sequence flowshop scheduling problem. Eur. J. Oper. Res. 177(3), 2033–2049 (2007)
28. Saskena, J.: Mathematical model of scheduling clients through welfare agencies. Journal of the Canadian Operational Research Society 8, 185–200 (1970)
29. Silberholz, J., Golden, B.: The generalized traveling salesman problem: A new genetic algorithm approach. In: Edward, K.B., et al. (eds.) Extending the horizons: Advances in Computing, Optimization and Decision Technologies, vol. 37, pp. 165–181. Springer, Heidelberg (1997)

30. Snyder, L., Daskin, M.: A random-key genetic algorithm for the generalized traveling salesman problem. Eur. J. Oper. Res. 174, 38–53 (2006)
31. Storn, R., Price, K.: Differential evolution - a simple and efficient adaptive scheme for global optimization over continuous spaces. ICSI, Technical Report TR-95-012 (1995)
32. Srivastava, S., Kumar, S., Garg, R., Sen, R.: Generalized traveling salesman problem through n sets of nodes. Journal of the Canadian Operational Research Society 7, 97–101 (1970)
33. Tasgetiren, M., Sevkli, M., Liang, Y.-C., Gencyilmaz, G.: Particle Swarm Optimization Algorithm for the Single Machine Total Weighted Tardiness Problem. In: The Proceeding of the World Congress on Evolutionary Computation, CEC 2004, pp. 1412–1419 (2004)
34. Tasgetiren, M., Suganthan, P., Pan, Q.-K.: A discrete particle swarm opti-mization algorithm for the generalized traveling salesman problem. In: The Proceedings of the 9th annual conference on genetic and evolutionary computation (GECCO 2007), London UK, pp. 158–167 (2007)
35. Tasgetiren, M., Suganthan, P., Pan, Q.-K., Liang, Y.-C.: A genetic algorithm for the generalized traveling salesman problem. In: The Proceeding of the World Congress on Evolutionary Computation (CEC 2007), Singapore, pp. 2382–2389 (2007)
36. Tasgetiren, M., Pan, Q.-K., Suganthan, P., Chen, A.: A hybrid iterated greedy algorithm for the generalized traveling salesman problem, Computers and Industrial Engineering (submitted, 2008)

7
Discrete Set Handling

Ivan Zelinka

Tomas Bata Univerzity in Zlin, Faculty of Applied Informatics,
Nad Stranemi 4511, Zlin 76001, Czech Republic
zelinka@fai.utb.cz

Abstract. Discrete Set Handling and its application to permutative problems is presented in this chapter. Discrete Set is applied to Differential Evolution Algorithm, in order to enable it to solve strict-sence combinatorial problems. In addition to the theoretical framework and description, benchmark Flow Shop Scheduling and Traveling Salesman Problems are solved. The results are compared with published literature to illustrate the effectiveness of the developed approach. Also, general applications of Discrete Set Handling to Chaotic, non-linear and symbolic regression systems are given.

7.1 Introduction

In recent years, a broad class of algorithms has been developed for stochastic optimization, i.e. for optimizing systems where the functional relationship between the independent input variables x and output (objective function) y of a system S is not known. Using stochastic optimization algorithms such as Genetic Algorithms (GA), Simulated Annealing (SA) and Differential Evolution (DE), a system is confronted with a random input vector and its response is measured. This response is then used by the algorithm to tune the input vector in such a way that the system produces the desired output or target value in an iterative process.

Most engineering problems can be defined as optimization problems, e.g. the finding of an optimal trajectory for a robot arm, the optimal thickness of steel in pressure vessels, the optimal set of parameters for controllers, optimal relations or fuzzy sets in fuzzy models, etc. Solutions to such problems are usually difficult to find, since their parameters usually include variables of different types, such as floating point or integer variables. Evolutionary algorithms (EAs), such as the Genetic Algorithms and Differential Evolutionary Algorithms, have been successfully used in the past for these engineering problems, because they can offer solutions to almost any problem in a simplified manner: they are able to handle optimizing tasks with mixed variables, including the appropriate constraints, and they do not rely on the existence of derivatives or auxiliary information about the system, e.g. its transfer function.

Evolutionary algorithms work on populations of candidate solutions that are evolved in generations in which only the best−suited − or fittest − individuals are likely to survive. This article introduces Differential Evolution, a well known stochastic optimization algorithm. It explains the principles of permutation optimization behind DE and demonstrates how this algorithm can assist in solving of various permutation optimization problems.

Differential Evolution, which can also works on a population of individuals, is based on a few simple arithmetic operations. Individuals are generated by means of a few randomly selected individuals.

In the following text the principle of the DE algorithm and permutative optimization will be explained. The description is divided into short sections to increase the understandability of principles of DE and permutative optimization.

7.2 Permutative Optimization

A permutative optimization problem is one, where the solution representation is ordered and discrete; implying that all the values in the solutions are firstly *unique*, and secondly *concrete*.

In the general sense, if a problem representation is given as n, then the solution representation is always given as some combination of range $\{1,....,n\}$. For example, given a problem of size 4, the solution representation is $\{1,2,3,4\}$ and all its possible permutative combinations.

Two of the problems solved in this chapter, which are of this nature, are the Traveling Salesman Problem (TSP) and Flow Shop Scheduling (FSS) Problems as discussed in the following sections. The third subsection describes the 2 Opt Local search, which is a routine embedded in this heuristic to find better solutions within the neighbourhood of a solution.

7.2.1 Travelling Salesman Problem

A TSP is a classical combinatorial optimization problem. Simply stated, the objective of a travelling salesman is to move from city to city, visiting each city only once and returning back to the starting city. This is called a *tour* of the salesman. In mathematical formulation, there is a group of distinct cities $\{C_1, C_2, C_3, ..., C_N\}$, and there is given for each pair of city $\{C_i, C_j\}$ a distance $d\{C_i, C_j\}$. The objective then is to find an ordering π of cities such that the total time for the salesman is minimised. The lowest possible time is termed the optimal time. The objective function is given as:

$$\sum_{i=1}^{N-1} d\left(C_{\pi(i)}, C_{\pi(i+1)}\right) + d\left(C_{\pi(N)}, C_{\pi(1)}\right) \tag{7.1}$$

This quality is known as the *tour length*. Two branches of this problem exist, symmetric and asymmetric. A symmetric problem is one where the distance between two cities is identical, given as: $d\{C_i, C_j\} = d\{C_j, C_i\}$ for $1 \leq i, j \leq N$ and the asymmetric is where the distances are not equal. An asymmetric problem is generally more difficult to solve.

The TSP has many real world applications; VSLA fabrication [7] to X-ray crystallography [1]. Another consideration is that TSP is *NP-Hard* as shown by [12], and so any algorithm for finding optimal tours must have a worst-case running time that grows faster than any polynomial (assuming the widely believed conjecture that $P \neq NP$).

TSP has been solved to such an extent that traditional heuristics are able to find good solutions to merely a small percentage error. It is normal for the simple 3-Opt heuristic typically getting with 3-4% to the optimal and the *variable-opt* algorithm of [20] typically getting around 1-2%.

The objective for new emerging evolutionary systems is to find a guided approach to TSP and leave simple local search heuristics to find better local regions, as is the case for this chapter.

7.2.2 Flow Shop Scheduling Problem

A flow shop is a scheduling problem, typical for a manufacturing floor. The terminology for this problem is typical of a manufacturing sector. Consider n number of jobs $j(i = 1, ... n)$, and a number of machines M: $M(j = 1, ..., m)$.

A job consists of m operation and the jth of each job must be processed on machine j. So, one job can start on machine j if it is completed on machine j-1 and if machine j is free. Each job has a known processing time $p_{i,j}$. The operating sequence of the jobs is the same on all the machines. If one job is at the ith position on machine 1, then it will be on the ith position on all machines.

The objective function is then to find the minimal time for the completion of all the jobs on all the machines. A job J_i is a sequence of operations, having one operation for each of the M machines.

1. $J_i = \{O_{i1}, O_{i2}, O_{i3}, .., O_{iM}\}$, where O_{ij} represents the jth operation on J_i.
2. O_{ij} operation must be processed on M_j machine.
3. for each operation O_{ij}, there is a processing time $p_{i,j}$.

Now let a permutation be represented as $\{\Pi_1, \Pi_2, ..., \Pi_N\}$. The formulation of the completion time for $C(\Pi_i, j)$, for the ith job on the jth machine can be given as:

$$\begin{aligned} C(\Pi_1, 1) &= p_{\Pi_1, 1} & & \\ C(\Pi_1, 1) &= C(\Pi_{i-1}, 1) + p_{\Pi_1, 1}, & i &= 2, ..., N \\ C(\Pi_1, j) &= C(\Pi_1, j-1) + p_{\Pi_1, j}, & i &= 2, ..., M \\ C(\Pi_1, j) &= \max\{C(\Pi_{i-1}, 1), C(\Pi_1, j-1)\} + p_{\Pi_1, j}, & i &= 2, ..., N; j = 2, .., M \end{aligned} \quad (7.2)$$

The makespan or the completion time is given as the $C(\Pi_N, M)$, as the completion time of the *last* job in the schedule on the *last* machine.

7.2.3 2 Opt Local Search

A local search heuristic is usually based on simple tour modifications (exchange heuristics). Usually these are specified in terms of the class of operators (exchanges/moves), which is used to modify one tour into another. This usually works on a feasible tour, where a *neighborhood* is all moves, which can be reached, in a single move. The tour iterates till a better tour is reached.

Among simple local search algorithms, the most famous are 2–Opt and 3–Opt. The 2-Opt algorithm was initially proposed by [2] although it was already suggested by [5]. This move deletes two edges, thus breaking the tour into two paths, and then reconnects those paths in the other possible way as given in Fig 7.1.

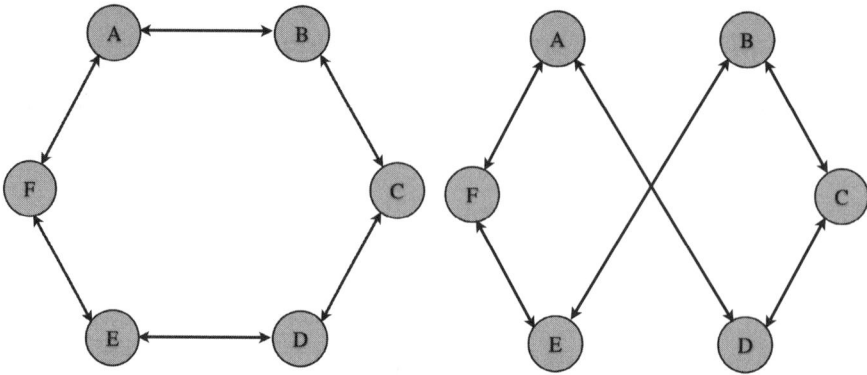

Fig. 7.1. 2-Opt exchange tour

7.3 Discrete Set Handling and Its Application

7.3.1 Introduction and Principle

In its canonical form, DE is only capable of handling continuous variables. However, extending it for optimization of integer variables is rather easy. Only a couple of simple modifications are required. First, for evaluation of the cost-function, integer values should be used. Despite this, the DE algorithm itself may still work internally with continuous floating-point values. Thus,

$$f_{\text{cost}}(y_i) \quad i = 1,..,n_{param}$$
$$\text{where}:$$
$$y_i = \begin{cases} x_i & \text{for continuous variables} \\ INT(x_i) & \text{for integer variables} \end{cases} \quad (7.3)$$
$$x_i \in X$$

INT() is a function for converting a real value to an integer value by truncation. Truncation is performed here only for purposes of cost function value evaluation. Truncated values are not assigned elsewhere. Thus, EA works with a population of continuous variables regardless of the corresponding object variable type. This is essential for maintaining the diversity of the population and the robustness of the algorithm.

Secondly, in case of integer variables, the population should be initialized as follows:

$$P^{(0)} = x_{i,j}^{(0)} = r_{i,j}\left(x_j^{(High)} - x_j^{(Low)} + 1\right) + x_j^{(Low)} \quad (7.4)$$
$$i = 1,...,n_{pop}, \quad j = 1,...,n_{param}$$

Additionally, the boundary constraint handling for integer variables should be performed as follows:

$$x_{i,j}^{(ML+1)} = \begin{cases} r_{i,j}\left(x_j^{(High)} - x_j^{(Low)} + 1\right) + x_j^{(Low)} \\ \quad if \quad INT\left(x_{i,j}^{(ML+1)}\right) < x_j^{(Low)} \vee INT\left(x_{i,j}^{(ML+1)}\right) > x_j^{(High)} \\ x_{i,j}^{(ML+1)} \quad \text{otherwise} \end{cases} \quad (7.5)$$

where,
$i = 1, ..., n_{pop}, \quad j = 1, ..., n_{param}$

Discrete values can also be handled in a straight forward manner. Suppose that the subset of discrete variables, $X(d)$, contains i elements that can be assigned to variable x:

$$X^{(d)} = x_i^{(d)} \quad i = 1,...,l \quad \text{where} \quad x_i^{(d)} < x_{i+1}^{(d)} \quad (7.6)$$

Instead of the discrete value x_i itself, its index, i, can be assigned to x. Now the discrete variable can be handled as an integer variable that is boundary constrained to range $\{1, 2, 3, .., N\}$. In order to evaluate the objective function, the discrete value, x_i, is used instead of its index i. In other words, instead of optimizing the value of the discrete variable directly, the value of its index i is optimized. Only during evaluation is the indicated discrete value used. Once the discrete problem has been converted into an integer one, the previously described methods for handling integer variables can be applied. The principle of discrete parameter handling is depicted in Fig 7.2.

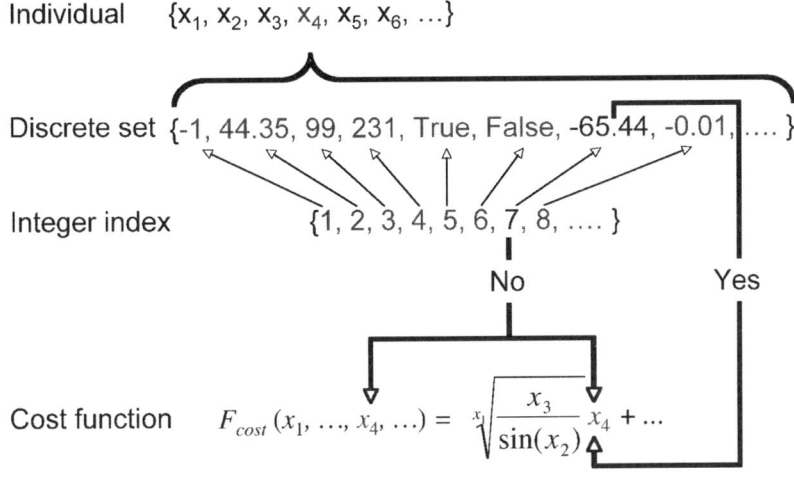

Fig. 7.2. Discrete parameter handling

7.3.2 DSH Applications on Standard Evolutionary Algorithms

DSH has been used in many previous experiments in standard EAs as well as in genetic programming like techniques. An example of the usage of DSH in mechanical engineering problem in C++ language in given in Fig 7.3.

Here, only the set of discrete values is described in order to show that DSH is basically a field of values (real values) and individuals in integer form serve like pointers to

```
// Mixed problem (Integer - Continuous - Discrete) - Case B !!!!!!!!!!!!!!!!!
// New Ideas in Optimization - Table 9 : Allowable spring steel wire diameters
// for the coil spring design problem
discrete[0] = 0.009; discrete[1] = 0.0095; discrete[2] = 0.0104; discrete[3] = 0.0118;
discrete[4] = 0.0128; discrete[5] = 0.0132; discrete[6] = 0.014; discrete[7] = 0.015;
discrete[8] = 0.0162; discrete[9] = 0.0173; discrete[10] = 0.018; discrete[11] = 0.020;
discrete[12] = 0.023; discrete[13] = 0.025; discrete[14] = 0.028; discrete[15] = 0.032;
discrete[16] = 0.035; discrete[17] = 0.041; discrete[18] = 0.047; discrete[19] = 0.054;
discrete[20] = 0.063; discrete[21] = 0.072; discrete[22] = 0.080; discrete[23] = 0.092;
discrete[24] = 0.105; discrete[25] = 0.120; discrete[26] = 0.135; discrete[27] = 0.148;
discrete[28] = 0.162; discrete[29] = 0.177; discrete[30] = 0.192; discrete[31] = 0.207;
discrete[32] = 0.225; discrete[33] = 0.244; discrete[34] = 0.263; discrete[35] = 0.283;
discrete[36] = 0.307; discrete[37] = 0.331; discrete[38] = 0.362; discrete[39] = 0.394;
discrete[40] = 0.4375; discrete[41] = 0.500;
```

Fig. 7.3. C++ DSH code

```
#include <stdlib.h>

int      tempval,MachineJob[25],Cmatrix[5][5];
int      loop1,loop2;

//in C language is [0][0] the first item of defined field i.e. [1][1] of
normaly defined matrix
MachineJob[0]=5;MachineJob[1]=7;MachineJob[2]=4;MachineJob[3]=3;MachineJob[4]=6;
MachineJob[5]=6;MachineJob[6]=5;MachineJob[7]=7;MachineJob[8]=6;MachineJob[9]=7;
MachineJob[10]=7;MachineJob[11]=8;MachineJob[12]=3;MachineJob[13]=8;MachineJob[14]=5;
MachineJob[15]=8;MachineJob[16]=6;MachineJob[17]=5;MachineJob[18]=5;MachineJob[19]=8;
MachineJob[20]=4;MachineJob[21]=4;MachineJob[22]=8;MachineJob[23]=7;MachineJob[24]=3;

for(loop1=0;loop1<25;loop1++)
   {
   tempval=MachineJob[loop1];
    MachineJob[loop1]=MachineJob[getIntPopulation(0,Individual)];
   MachineJob[getIntPopulation(0,Individual)]=tempval;
   };

//Competition time for all jobs on machine 1
Cmatrix[0][0]=MachineJob[0];
Cmatrix[0][1]=MachineJob[0]+MachineJob[1];
Cmatrix[0][2]=MachineJob[0]+MachineJob[1]+MachineJob[2];
Cmatrix[0][3]=MachineJob[0]+MachineJob[1]+MachineJob[2]+MachineJob[3];
Cmatrix[0][4]=MachineJob[0]+MachineJob[1]+MachineJob[2]+MachineJob[3]+MachineJob[4];

//Competition time jobs 1 on all machines
Cmatrix[1][0]=MachineJob[0]+MachineJob[5];
Cmatrix[2][0]=MachineJob[0]+MachineJob[5]+MachineJob[10];
Cmatrix[3][0]=MachineJob[0]+MachineJob[5]+MachineJob[10]+MachineJob[15];
Cmatrix[4][0]=MachineJob[0]+MachineJob[5]+MachineJob[10]+MachineJob[15]+
              MachineJob[20];

for(loop1=1;loop1<5;loop1++)
    for(loop2=1;loop2<5;loop2++)
         Cmatrix[loop1][loop2]=max(Cmatrix[loop1-1][loop2],Cmatrix[loop1][loop2-1])+
         MachineJob[5*loop1+loop2];

CostValue=Cmatrix[4][4];
```

Fig. 7.4. DSH FSS example

that field. A more complex example from FSS is now described in Fig 7.4. Discrete set has the name **MachineJob** and contains different values. Individuals again serve like an index.

More interesting applications of DSH can be found in genetic programming like techniques.

7.3.3 DSH Applications on Class of Genetic Programming Techniques

The term *symbolic regression* represents a process during which measured data sets are fitted such thereby a corresponding mathematical formula is obtained in an analytical way. An output of the symbolic expression could be, for example, $x^2 + y^3/K$, and the like. For a long time, symbolic regression was a domain of human calculations but in the last few decades it involves computers for symbolic computation as well.

The initial idea of symbolic regression by means of a computer program was proposed in Genetic Programming (GP) [8, 9]. The other approaches are Grammatical Evolution (GE) developed in [19, 13] and Analytic Programming (AP) in [27]. Oher interesting investigations using symbolic regression were carried out in [6] on Artificial Immune Systems and Probabilistic Incremental Program Evolution (PIPE), which generates functional programs from an adaptive probability distribution over all possible programs. As an extension of GE to the another algorithms is also [14], where DE was used with the GE. Symbolic regression, generally speaking, is a process which combines, evaluates and creates more complex structures based on some elementary and noncomplex objects, in an evolutionary way. Such elementary objects are usually simple mathematical operators $(+, -, *, ...)$, simple functions (sin, cos, And, Not,.), user-defined functions (simple commands for robots – MoveLeft, TurnRight,.), etc.

An output of symbolic regression is a more complex *object* (formula, function, command,.), solving a given problem like data fitting of the so-called Sextic and Quintic problem described by Equation 7.7) [10, 26], randomly synthesized function by Equation 7.8 [26], Boolean problems of parity and symmetry solution (basically logical circuits synthesis) by Equation 7.9) [11, 27], synthesis of Chaos by utilizing DSH and Evolutionary Algorithms [28] given in Table 7.1 and in Figs 7.7 – 7.10.

Synthesis of quite complex robot control command by Equation 7.10 [10, 15] is also accomplished with DSH. Equation 7.7 – 7.10 mentioned are just a few samples from numerous repeated experiments done by AP, which are used to demonstrate how complex structures can be produced by symbolic regression in general for different problems.

$$x\left(K_1 + \frac{(x^2 K_3)}{K_4(K_5 + K_6)}\right) \bullet (-1 + K_2 + 2x(-x - K_7)) \tag{7.7}$$

$$\sqrt{t}\left(\frac{1}{\log(t)}\right)^{\sec^{-1}(1.28)} \log^{\sec^{-1}(1.28)}(\sinh(\sec(\cos(1)))) \tag{7.8}$$

```
Nor[ (Nand[Nand[B || B, B && A], B]) && C && A && B,
    Nor[(! C && B && A || ! A && C && B || ! C && ! B && ! A) &&
        (! C && B && A || ! A && C && B || ! C && ! B && ! A) ||
    A && (! C && B && A || ! A && C && B || ! C && ! B && ! A),
    (C || ! C && B && A || ! A && C && B || ! C && ! B && ! A) && A]]
```
$$\tag{7.9}$$

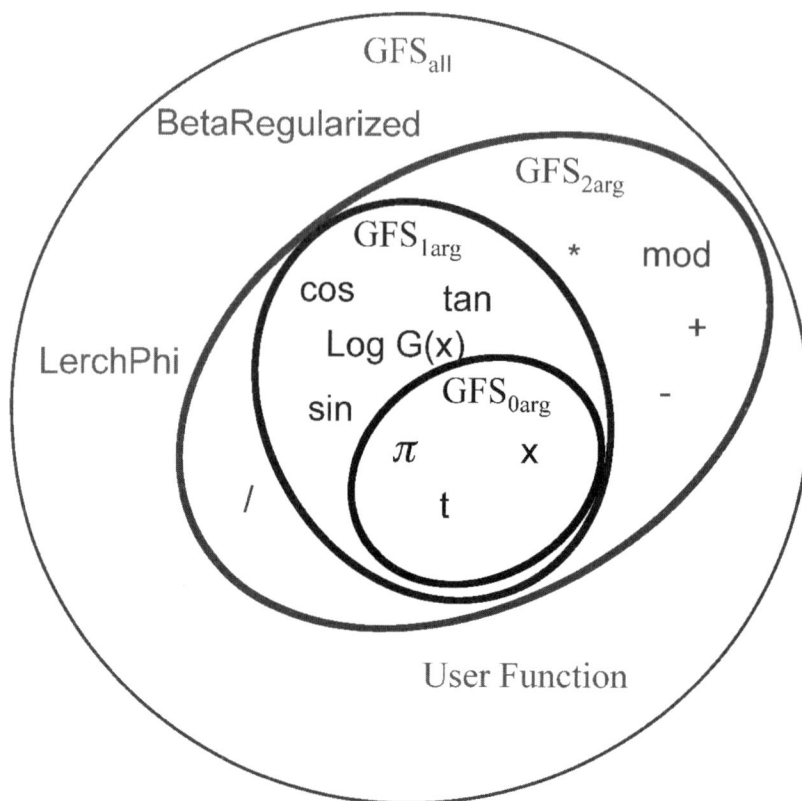

Fig. 7.5. Principle of the general functional set

```
TreeForm[IfFoodAhead[Move, Prog3[IfFoodAhead[Move, Right],
    Prog2[Right, Prog2[Left, Right]],
    Prog2[IfFoodAhead[Move, Left], Move]]]]
```

(7.10)

The final method described here and used for experiments is called Analytic Programming (AP), which has been compared to GP with very good results (see, for example, [26, 15, 27]) or visit the online univeristy website [http://www.fai.utb.cz/people/zelinka/ap].

The basic principles of AP were developed in 2001 and first published in [24, 25]. AP is also based on the set of functions, operators and terminals, which are usually constants or independent variables alike, for example:

1. functions: sin, tan, tanh, And, Or
2. operators: +, -, *, /, dt,
3. terminals: 2.73, 3.14, t,

All these *mathematical* objects create a set, from which AP tries to synthesize an appropriate solution. Because of the variability of the content of this set, it is called a

Individual parameters {1, 6, 7, 8, 9, 9} are used by AP like pointers into GFS and through serie of mappings m1 - m5 final formula $\sin(\tan(t)) + \cos(t)$ is created.

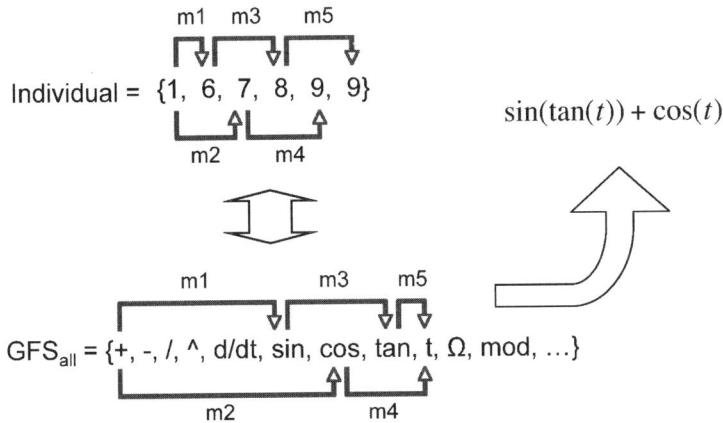

Fig. 7.6. Main principles of AP based on DSH

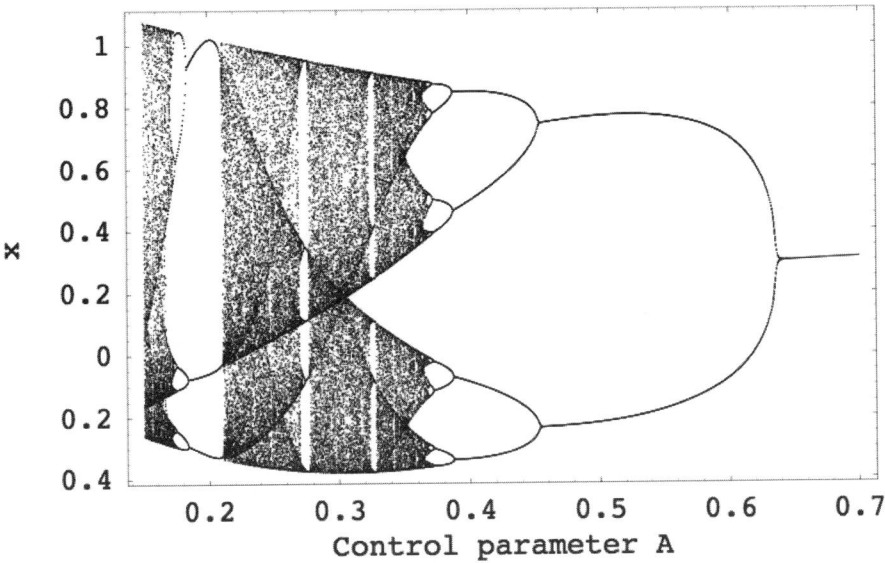

Fig. 7.7. Bifurcation diagram, exhibiting chaos and generated by artificially synthesied equations

general functional set (GFS). The structure of GFS is nested, i.e., it is created by subsets of functions according to the number of their arguments (The content of GFS is dependent only on the user. Various functions and terminals can be mixed together. For example, GFS_{all} is a set of all functions, operators and terminals, GFS_{3arg} is a subset containing functions with maximally three arguments, GFS_{0arg} represents only terminals, etc. (see Fig 7.5).

Table 7.1. Selected solutions synthesized by EA and DSH

Equation	Bifurcations and chaos
$A - x \left(-A + \dfrac{x\left(-\frac{A}{x} + x + Ax\right)}{A + x^2} \right)$	$\{0.4\}$
$\dfrac{A\left(2A - 2x^2 - 3x(A - x - Ax)\right)}{-A + x - x^2}$	$\{0.1, 0.13\}$ $\{0.8, 1.2\}$
$-x - \dfrac{1 - 2A + 2x + 2A^2 x}{1 - A + \frac{A^2 - x}{x} + x}$	$\{0.3, 0.5\}$
$\dfrac{x - A\left(A - x - 2x^2\right)}{-A - x + Ax^2 - A\left(-A + \frac{A^3}{x} + 2x\right)}$	$\{0.4, 0.5\}$
$\dfrac{2A(-2A + 2x)}{x + \frac{1 + A^2 + x}{x}}$	$\{0.4\}$
$\dfrac{x}{(3A + 2x)\left(-1 - A - x + \frac{x(A + 2x)}{A^2 + x}\right)}$	$\{0.12, 0.23\}$ $\{0.3, 0.36\}$

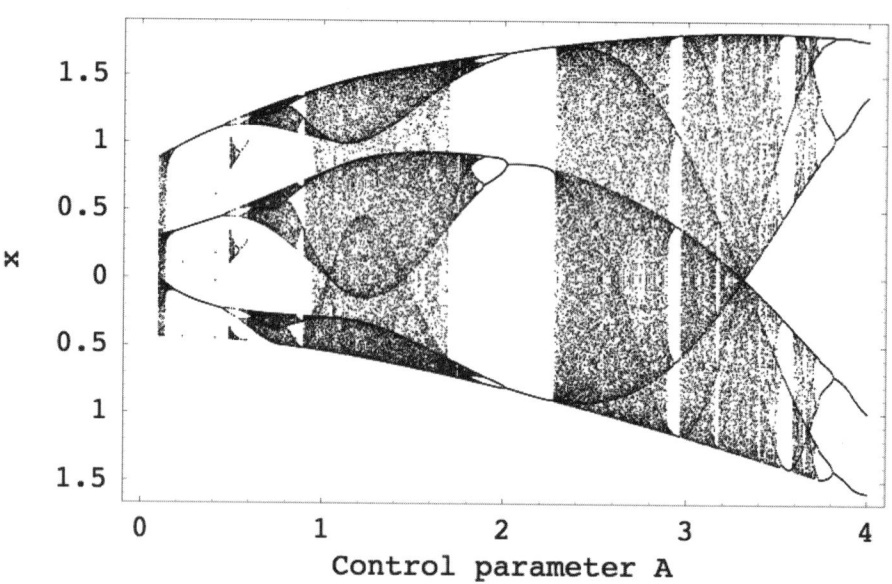

Fig. 7.8. Another bifurcation diagram

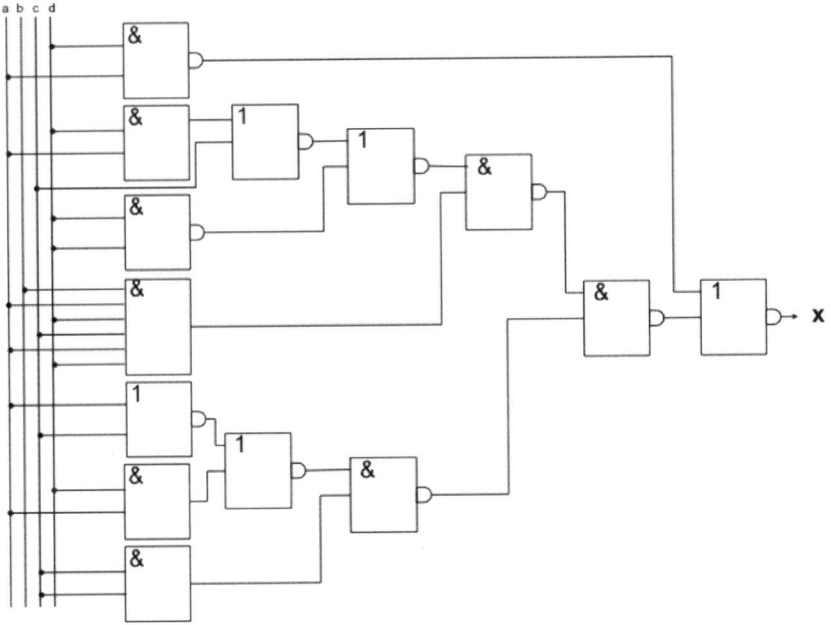

Fig. 7.9. Synthesized logical circuit by means of EA and DSH

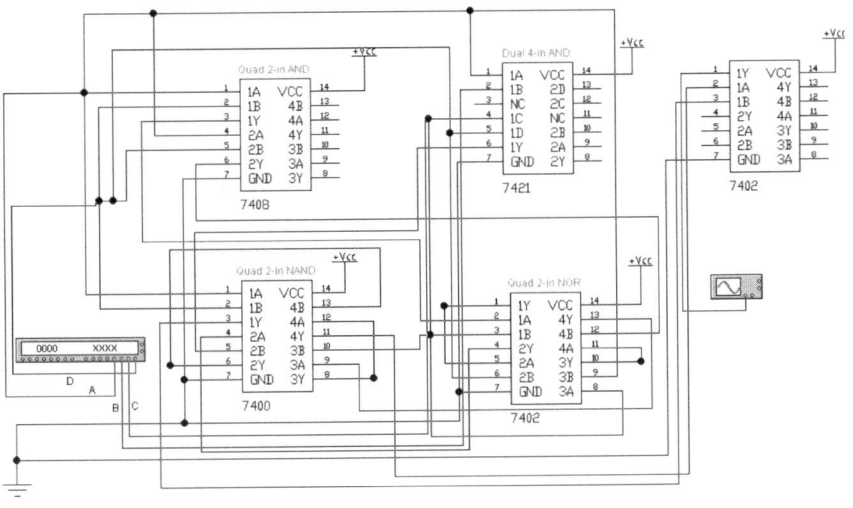

Fig. 7.10. Realization of logical circuit from Equation 7.7

AP, is a mapping from set of individuals into set of posssible programs. Individuals in population and used by AP consist of non-numerical expressions (operators, functions, .), as described above, which are in the evolutionary process represented by their integer position indexes (Fig 7.6). This index then serves as a pointer into the set of expressions and AP uses it to synthesize the resulting function-program for cost function evaluation.

AP was evaluated in three versions. These three versions utilize for program synthesis the same set of functions, terminals, etc., as in GP [9, 10]. The second version labelled as AP_{meta} (the first version, AP_{basic}) is modified in the sense of constant estimation. For example, the so-called sextic problem was used in [9] to randomly generate constants, whereas AP here uses only one, called K, which is inserted into Equation 7.11 below at various places by the evolutionary process. When a program is synthesized, all Ks are indexed as $K_1, K_2, \ldots, K_n$ to obtain Equation 7.12 in the formula, and then all K_n are estimated by using a second evolutionary algorithm, the result of which can be, for example, Equation 7.13. Because EA (slave) *works under* EA (master), i.e., $EA_{master} \rightarrow$ program $\rightarrow$ K indexing $\rightarrow EA_{slave} \rightarrow$ estimation of K_n, this version is called AP with metaevolution, denoted as AP_{meta}.

$$\frac{x^2 + K}{\pi^K} \tag{7.11}$$

$$\frac{x^2 + K_1}{\pi^{K_2}} \tag{7.12}$$

$$\frac{x^2 + 3.56}{\pi^{-229}} \tag{7.13}$$

Because this version is quite time-consuming, AP_{meta} was further modified to the third version, which differs from the second one in the estimation of K. This is accomplished by using a suitable method for nonlinear fitting (denoted AP_{nf}). This method has shown the most promising performance when unknown constants are present. Results of some comparative simulations can be found in [25, 26, 27].

7.4 Differential Evolution in Mathematica Code

Differential Evolution used in all experiments reported in this chapter has been based on the Mathematica Programming environment. The aim of this part is to describe the structure of the DE code and final code development. Source codes reported here are only a part of fully developed notebook in environment Mathematica. Only the main ideas and some parts of the final code are described here.

For the beginning of DE code development, it is important to prepare the population and set all DE algorithm parameters like F, CR, NP and Generation. Population is initialized as shown in Fig 7.11.

```
In[25]:=  Population = DoPopulation[NP, Specimen]

Out[25]=  {{534.695, {-442.422, -188.47}}, {191.21, {194.845, -437.931}},
          {-70.135, {-127.976, 99.3825}}, {-208.07, {214.324, 244.138}},
          {-41.6243, {-236.027, -128.204}}, {161.461, {335.943, 355.91}},
          {106.047, {-317.752, -402.607}}, {-157.266, {-119.503, 163.852}},
          {464.407, {507.525, 502.251}}, {-62.8734, {160.401, -149.99}}}
```

Fig. 7.11. Population initialization

In[26]:= `Table[Random[Integer, {1, NP}], {i, 3}]`

Out[26]= `{3, 6, 8}`

Fig. 7.12. Random values

```
SelectOther[active_] := Module[{},
  rand = {0, 0, 0};
  While[rand[[1]] == rand[[2]] || rand[[1]] == rand[[3]] || rand[[2]]
    == rand[[3]] || active == rand[[1]] || active == rand[[2]] ||
    active == rand[[3]], rand = Table[Random[Integer, {1, NP}], {i, 3}]];
  Return[rand]
]
```

Fig. 7.13. SelectOther function

```
SelectOther[active_] := Module[{rand = {0, 0, 0, 0, 0}, allvals},
  While[allvals = Append[rand, active];
   Length[allvals] != Length[Union[allvals]],
   rand = Table[Random[Integer, {1, NP}], {i, 5}]];
  Return[rand]]
```

Fig. 7.14. SelectOther compressed function

In[29]:= `SelectOther[1]`

Out[29]= `{2, 8, 3}`

Fig. 7.15. Three random indexes

This command returns the initial population of individuals with the structure Cost-Value, parameter1, parameter2, . . , parameterNP. *NP* is a size of the population. Canonical version of the DE is based on the selection of the three (or more based on DE version) randomly chosen individuals from the population. Random selection, or more precisely, random selection of three pointers, can be done by command *Table* in Fig 7.12.

The random values selects pointers to three individuals of *NP*. To avoid the possibility that two or more will be the same, *SelectOther* function is used. Its argument *active* is a pointer to the actively selected individual − parent. *SelectOther* function is shown in Fig 7.13.

SelectOther function can also be given in a compressed form as in Fig 7.14.

Counters $\{1, 2, .., NP\}$ and $\{i, 3\}$ are used for selection of three different individuals from NP individuals. Fig 7.15 shows three individuals selected from the first solution (parent). Note that all these individuals differ from the first one (position 3).

Till this point, the initial population has been initialized and three individuals have been randomly selected from the population. In the following step, the function *SelectOther* is applied to the entire population at once. Mathematica language allows *parallel−like* programming, which is visible throughout of the code. This is also the case of the following command of function *MapIndexed* which is used to apply *SelectOther* on all individuals, so that the *virtual* population of pointers (randomly selected

In[30]:= `TRVIndex = MapIndexed[SelectOther[#2[[1]]] &, Population]`

Out[30]= {{7, 9, 8}, {5, 1, 7}, {6, 2, 10}, {10, 7, 6},
{10, 6, 4}, {8, 7, 5}, {1, 2, 4}, {1, 7, 9}, {5, 3, 2}, {2, 9, 8}}

Fig. 7.16. TVR Index of pointers

In[31]:= `TRV = Population[[#1]] & /@ TRVIndex`

Out[31]= {{{106.047, {-317.752, -402.607}}, {464.407, {507.525, 502.251}},
{-157.266, {-119.503, 163.852}}}, {{-41.6243, {-236.027, -128.204}},
{534.695, {-442.422, -188.47}}, {106.047, {-317.752, -402.607}}},
{{161.461, {335.943, 355.91}}, {191.21, {194.845, -437.931}},
{-62.8734, {160.401, -149.99}}}, {{-62.8734, {160.401, -149.99}},
{106.047, {-317.752, -402.607}}, {161.461, {335.943, 355.91}}},
{{-62.8734, {160.401, -149.99}}, {161.461, {335.943, 355.91}},
{-208.07, {214.324, 244.138}}}, {{-157.266, {-119.503, 163.852}},
{106.047, {-317.752, -402.607}}, {-41.6243, {-236.027, -128.204}}},
{{534.695, {-442.422, -188.47}}, {191.21, {194.845, -437.931}},
{-208.07, {214.324, 244.138}}}, {{534.695, {-442.422, -188.47}},
{106.047, {-317.752, -402.607}}, {464.407, {507.525, 502.251}}},
{{-41.6243, {-236.027, -128.204}}, {-70.135, {-127.976, 99.3825}},
{191.21, {194.845, -437.931}}}, {{191.21, {194.845, -437.931}},
{464.407, {507.525, 502.251}}, {-157.266, {-119.503, 163.852}}}}

Fig. 7.17. TVR Index of pointers for entire population

triplets) is created. Fig 7.16 shows the varible *TRVIndex* (trial vector index) created from the *MapIndexed* function.

To unfold the code, the operator /@ (function *Map*) is used, which takes all the arguments of pointers shown in Fig 7.16 and creates an entire array of pointers for the population, given in Fig 7.17.

The result of Fig 7.17 is a list of physically selected individuals (three for each parent). Application of mutation principle and all DE arithmetic operations on TRV list is straight forward. It is accomplished by the means of the operator /@ which in this case applies the arithmetic operation from the left to the elements of the TRV list. Entity #1[[X, 2]] in the arithmetic formula F∗ (#1[[1,2]] − #1[[2,2]]) +#1[[3,2]] represents X_{th} individual from the selected triplets in TRV. The *Noisy* vector is thus calculated like in Fig 7.18.

`Noisy = F * (#1[[1, 2]] - #1[[2, 2]]) + #1[[3, 2]] & /@ TRV`

{{-779.725, -560.034}, {-152.636, -354.393}, {273.28, 485.082},
{718.466, 558.003}, {73.8901, -160.582}, {-77.4278, 324.963},
{-295.49, 443.706}, {407.789, 673.56}, {108.404, -620.}, {-369.647, -588.294}}

Fig. 7.18. Noisy Vector

The output of Fig 7.18 is a set of *Noisy* vectors (cardinality of NP), which is consequently used to generate trial vectors − individuals. Parameter selection from the parent or noisy vector is done by the condition If[Cr < Random[]...]. *Flatten* is only a cosmetic command which removes redundant brackets, generated by the command *Table*. In the standard programming approach the command *For* would be used. To

```
Trial = Flatten[Table[If[Cr < Random[],
        Population[[i, 2, j]], #1[[i, j]]],
        {i, NP}, {j, Dim}] & /@ {Noisy}, 1]
```

{{-442.422, -560.034}, {194.845, -354.393}, {-127.976, 99.3825},
 {718.466, 244.138}, {-236.027, -128.204}, {335.943, 355.91},
 {-317.752, -402.607}, {-119.503, 163.852}, {507.525, 502.251}, {160.401, -149.99}}

Fig. 7.19. Trial Vector

```
BoundaryChecking = Flatten[MapIndexed
        [CheckInterval[#1, #2] &, #1]
        , 1] & /@ (Trial)
```

{{-442.422, -2.47524}, {194.845, -354.393}, {-127.976, 99.3825},
 {300.954, 244.138}, {-236.027, -128.204}, {335.943, 355.91},
 {-317.752, -402.607}, {-119.503, 163.852}, {507.525, 502.251}, {160.401, -149.99}}

Fig. 7.20. Boundary Checking

In[44]:= `IndividualsCostValue = {CostFunction[#1], #1} & /@ (BoundaryChecking)`

Out[44]= {{364.224, {-442.422, -2.47524}}, {-200.331, {194.845, -354.393}},
 {-70.135, {-127.976, 99.3825}}, {279.978, {300.954, 244.138}},
 {-41.6243, {-236.027, -128.204}}, {161.461, {335.943, 355.91}},
 {106.047, {-317.752, -402.607}}, {-157.266, {-119.503, 163.852}},
 {464.407, {507.525, 502.251}}, {-62.8734, {160.401, -149.99}}}

Fig. 7.21. Individual Cost Value

```
NewPopulation = MapThread[If[#1[[1]] < #2[[1]], #1, #2] &,
        {Population, (IndividualsCostValue)}]
```

{{364.224, {-442.422, -2.47524}}, {-200.331, {194.845, -354.393}},
 {-70.135, {-127.976, 99.3825}}, {-208.07, {214.324, 244.138}},
 {-41.6243, {-236.027, -128.204}}, {161.461, {335.943, 355.91}},
 {106.047, {-317.752, -402.607}}, {-157.266, {-119.503, 163.852}},
 {464.407, {507.525, 502.251}}, {-62.8734, {160.401, -149.99}}}

Fig. 7.22. Next Population Selection

avoid setting of local or global variables for the trial vector list, the *Table* command is used instead of *For*. *Trial* vectors are returned in the list given in Fig 7.19, which is created automatically.

All the *Trial* vectors are created at once. Before the fitness is calculated, the population of the trial individuals are checked for boundary conditions. If some parameter is out of the allowed boundary, then it is *randomly* returned back. The function is given in Fig 7.20.

Now, there exists a repaired set of trial vectors, which is evaluated by the cost function. It is done by the function *CostFunction* applied by /@ on the *BoundaryChecking* set. Note that the body of each individual in Fig 7.21 is enlarged by the function *IndividualsCostValue*.

The better individual of both *parent* and *child* is selected into the new population by means of the *MapThread* function.

```
NewPop[Pop_] := Module[{},
  TRVIndex = MapIndexed[SelectOther[#2[[1]]] &, Population];
  TRV = Population[[#1]] & /@ TRVIndex;
  Noisy = F * (#1[[1, 2]] - #1[[2, 2]]) + #1[[3, 2]] & /@ TRV;
  Trial = Flatten[Table[If[Cr < Random[], Pop[[i, 2, j]], #1[[i, j]]],
       {i, NP}, {j, Dim}] & /@ {Noisy}, 1];
  BoundaryChecking = Flatten[MapIndexed[CheckInterval[#1, #2] &, #1], 1]
       & /@ (Trial);
  IndividualsCostValue = {CostFunction[#1], #1} & /@ (BoundaryChecking);
  NewPopulation = MapThread[If[#1[[1]] < #2[[1]], #1, #2] &,
       {Population, (IndividualsCostValue)}]
]
```

Fig. 7.23. Compiled DE crossover code

In[47]:= `np = NewPop[Population];`
`MatrixForm[np]`

Out[48]//MatrixForm=

$$\begin{pmatrix} -1.65327 & \{5, 11.2\} \\ 86.0232 & \{-297.175, -437.931\} \\ -70.135 & \{-127.976, 99.3825\} \\ -208.07 & \{214.324, 244.138\} \\ -41.6243 & \{-236.027, -128.204\} \\ 161.461 & \{335.943, 355.91\} \\ -341.433 & \{-297.175, -150.458\} \\ -216.653 & \{-499.82, 163.852\} \\ 464.407 & \{507.525, 502.251\} \\ -62.8734 & \{160.401, -149.99\} \end{pmatrix}$$

Fig. 7.24. Function call of New population

If all the preceding steps are joined together, then final DE code in Mathematica is given in Fig 7.23.

When the function *NewPop* in Fig 7.23 is called with the variable *Population* like an argument the new population is created as shown in Fig 7.24.

When the output of *NewPop* in Fig 7.24 is repeatedly used as an input in some loop procedure (one loop – one generation), the *DE* algorithm is iterated.

Some additive procedures can also be used, like selection of the best individual from the population. An example is given in Fig 7.25.

A more compressed (but less readable) and similar version of DE is shown in Fig 7.26.

Canonical version of the DE described is a priori suitable for the real valued variables. However, due to the problems being solved here are based on integer–valued variables and permutative problems, some additional subroutines have been added to the DE code. The first one is a *Repair* subroutine. An input of this subroutine is an infesible solution and the output is a repaired solution so that each variable only appears once in the solution.The routine is shown in Fig 7.27.

7 Discrete Set Handling

```
ExpForm[nmbr_] := PaddedForm[nmbr, {6, 5}, ExponentFunction → (#1 &),
    NumberFormat → (#1 <> "<E>" <> #3 &), NumberSigns → {"-", "+"}];

BestInd[pop_] := Module[{best, ind, str},
    best = Position[{##}, Min[##]][[1]][[1]] & @@ Transpose[pop][[1]];
    ind = pop[[best]];
    str = "Best individual is on position " <> ToString[best] <>
      " with cost value " <> ToString[ExpForm[ind[[1]]]] <> " and
      parameters " <> ToString[ind[[2]]]];
    Print[str];
    Return[Flatten[{best, ind}, 1]]
    ]

BestInd[np]
```

```
Best individual is on position 7 with cost
   value -3.41433<E>2 and parameters {-297.175, -150.458}

  {7, -341.433, {-297.175, -150.458}}
```

Fig. 7.25. Best individual from population

```
NewPop[Pop_] := MapThread[If[#1[[1]] < #2[[1]]], #1, #2] &,
  {Pop, ({CostFunction[#1], #1} & /@ (Flatten
      [MapIndexed[CheckInterval[#1, #2] &, #1], 1] & /@
      (Flatten[Table[If[Cr < Random[], Pop[[i, 2, j]],
        #1[[i, j]]], {i, NP}, {j, Dim}] & /@ {F * (#1[[1, 2]]
          - #1[[2, 2]]) + #1[[3, 2]] & /@ (Pop[[#1]] & /@ MapIndexed
          [SelectOther[#2[[1]]] &, Pop])}, 1])))}]
```

Fig. 7.26. Compressed DE form

```
Repair[Sol_] := Module[{Temp, MissingValue, Solution, Pos, Size},
  Solution = Sol; Size = Length[Solution];
  MissingValue = RandomRelist[Complement[Range[Size], Solution]];
  Pos = Position[Solution, #] & /@ Range[Size];
  (Solution = Drop[Solution, {#}]) & /@ (Pos = Sort[Flatten[MapIndexed
      [Drop[#1, 1] &, #1] &[MapIndexed[RandomRelist[#1] &, #] &[
        Join[Temp[[#]]] & /@ (#1[[# & /@ Range[Length[#1]]]])) &[Flatten[
          Position[Flatten[Dimensions /@ (Temp = Flatten[Pos[[#]]] & /@
            Range[Size])], _?(1 < # &)]]]]], Greater]);
  Pos = Sort[Pos, Less];
  MapThread[(Solution = Insert[Solution, #1, #2]) &, {MissingValue, Pos}]
  Return[Solution]
  ]
```

Fig. 7.27. Repair routine

The *Repair* function is broken down and explained in-depth. The initial process is to find all the *missing values* in the solution. Since this is a *permutative* solution, each value is exist only *once* in the solution. Therefore it stands to reason that if there are more than one single value in the solution, then some values will be missing.

The function:
MissingValue = RandomRelist[Complement[Range[Size], Solution]]; finds the missing values in the solution.

The second phase is to map all the values in the solution. The routine:
Pos = Position[Solution, #]&/@Range[Size]; maps the *occurrence* of each value in the solution.

The *repetitive* values are identified in the function:
Flatten[Position[Flatten[Dimensions/@
    (Temp = Flatten[Pos[[#]]]&/@
        Range[Size])], _?(1 < #&)]]

The routine: Join[Temp[[#]]]&/@(#1[[#&/@Range[Length[#1]]]])& calculates the positions of the *replicated* values in the solution.

These replicated positions are *randomly shuffled*, since the objective is not to create any bias to replacement. This routine is given in the function:
MapIndexed[RandomRelist[#1]&, #]&

The variable *Pos* isolates the positions of replicated values which will be replaced as given in:
Pos = Sort[Flatten[MapIndexed[Drop[#1, 1]&, #1]&

The routine: Drop[Solution, {#}]&/@ removes the replicated values from the solution.

The final routine:
MapThread[(Solution = Insert[Solution, #1, #2])&, {MissingValue, Pos}]; inserts the missing values from the array *Missing Value* into randomly allocated indexes identified by variable *Pos*.

DE is consequently modified so that before the function *CostFunction* a *Repair/@DSH* function is used as in Fig 7.28.

```
DERand1Bin[Pop_] := MapThread[If[#1[[1]] < #2[[1]], #1, #2] &,
    {Pop, ({CostFunction[#1, Prob, Mach], #1} & /@
        (Repair/@DSH[Flatten[MapIndexed[CheckInterval[#1, #2] &,
            #1], 1] & /@ (Flatten[Table[If[Cr < Random[], Pop[[i, 2, j]],
                #1[[i, j]]], {i, NP}, {j, Dim}] & /@
            {F * (#1[[1, 2]] - #1[[2, 2]]) + #1[[3, 2]] & /@ (Pop[[#1]] & /@MapIndexed[
                SelectOtherRand1Bin[#2[[1]]] &, Pop])}, 1])
        ]
    ))}]
```

Fig. 7.28. Repair DSH routine

In[86]:= **DS = {M1, M2, M3, M4, M5, M6, M7, M8, M9, M10}**

Out[86]= {M1, M2, M3, M4, M5, M6, M7, M8, M9, M10}

Fig. 7.29. Discrete Set

In[87]:= **DSH[Pop_] := Module[{},**
        RoundPop = Round[Pop];
        DS[[#1]] & /@ #1 & /@ RoundPop
        ]

Fig. 7.30. Discrete Set

In[88]:= `DSH[BoundaryChecking] // MatrixForm`

Out[88]//MatrixForm=
$$\begin{pmatrix} M4 & M6 & M3 & M8 & M3 & M10 & M7 & M1 & M4 & M5 \\ M2 & M4 & M8 & M5 & M5 & M2 & M7 & M10 & M10 & M4 \\ M8 & M1 & M2 & M1 & M6 & M7 & M9 & M3 & M3 & M7 \\ M7 & M3 & M9 & M3 & M6 & M9 & M2 & M7 & M8 & M9 \\ M4 & M2 & M8 & M5 & M7 & M1 & M5 & M1 & M2 & M4 \\ M6 & M2 & M9 & M4 & M1 & M9 & M2 & M6 & M6 & M5 \\ M9 & M6 & M7 & M4 & M6 & M5 & M9 & M9 & M10 & M5 \\ M3 & M5 & M9 & M2 & M7 & M4 & M9 & M4 & M10 & M5 \\ M9 & M8 & M6 & M4 & M9 & M3 & M5 & M1 & M4 & M8 \\ M2 & M2 & M4 & M4 & M9 & M7 & M6 & M4 & M5 & M4 \end{pmatrix}$$

Fig. 7.31. Discrete Set Output

A discrete set can be created as shown in Fig 7.29.
The DSH function is given in Fig 7.30.
The result of applying the DSH set on the population is given in Fig 7.31.

Such or similar set can be used in other different methods (if needed) like fuzzy logic etc. Due to the nature of permutative problems, (sequence has to be complete and unique), the discrete set been set to the same sequence of numbers.

Due to the complex nature of permutative problems, a *Local Search* routine has been added to the heuristic. Local search is used to search in the *neighbourhood* of the current solutions. Keeping in mind the computational nature of the code, a 2 OPT local search outine was selected as in Fig 7.32.

```
LocalSearch[Sol_] :=
 Module[{Solution, NewSolution, CostVal, NewCostVal, Temp},
  CostVal = Sol[[1]]; Solution = Sol[[2]]; NewCostVal = CostVal;
      NewSolution = Solution;
  Label[start]; CostVal = NewCostVal; NewSolution = Solution;
  Do[
   Temp = Solution[[i]]; Solution[[i]] = Solution[[j]];
      Solution[[j]] = Temp;
   NewCostVal = CostFunction[Solution, Prob, Mach];
   If[NewCostVal < CostVal, Goto[start]],
      {i, Job - 1}, {j, i + 1, Job}];
  Solution = {CostVal, NewSolution}; Return[Solution]
 ]
```

Fig. 7.32. Local Search routine

The current fitness of the solution is kept in the variable *CostVal*, and the current active solution is kept in *Solution*. The start flag is *Label[start]*.

Two *iterators* are activated, i, which is the index to the current variable in the solution and j, which is the iterator from the current position indexed by i till the end of the solution given as $\{j, i+1, \text{Job}\}$.

Each two values in the solution are taken pairwise and exchanged as
Temp = Solution[[i]]; Solution[[i]] = Solution[[j]]; Solution[[j]] = Temp, where *Temp* is the intermediary placeholder. Another syntax for this process can be given as {Solution[[i]], Solution[[j]]} = {Solution[[j]], Solution[[i]]}. Each value indexed by *i* and *j* are exchanged.

The new fitness of the solution is calculated. If the new fitness is better than the old value, then the new solution is admitted into the population and the starting position is again set to *Label*[*start*] given as If[NewCostVal < CostVal, Goto[start]]. This process iterates till the index *i* iterates to the end of the solution {i, Job − 1} taking into account all the resets done by the finding of new solutions.

The outline of the entire code is given in Fig 7.33 and the data flow diagram is given in Fig 7.34.

1. Input : $D, G_{max}, NP \geq 4, F \in (0, 1+), CR \in [0, 1]$, initial bounds : $\mathbf{x}^{(lo)}, \mathbf{x}^{(hi)}$.
2. Initialize : *DoPopulation*[*NP*, *Specimen*]
3. While $G < G_{max}$

$\forall i \leq NP$

Create TRVIndex by command :
$TRVIndex = MapIndexed[SelectOther[\#2[[1]]]\&, Population]$

Selection of three vectors by TRVIndex
$TRV = Population[[\#1]]\&/@TRVIndex$

Create noisy vectors
$Noisy = F * (\#1[[1, 2]] - \#1[[2, 2]]) + \#1[[3, 2]]\&/@TRV$

Create trial vectors :
$Trial = Flatten[Table[If[Cr < Random[], Pop[[i, 2, j]], \#1[[i, j]]],$
$\{i, NP\}, \{j, Dim\}]\&/@\{Noisy\}, 1]$

Check for boundary :
$BoundaryChecking = Flatten[MapIndexed[CheckInterval[\#1, \#2]$
$\&, \#1], 1]\&/@(Trial)$

Cost value
$IndividualsCostValue = \{CostFunction[\#1], \#1\}\&/@(BoundaryChecking)$

DSH conversion :
$Repair/@DSH$

New population :
$NewPopulation = MapThread[If[\#1[[1]] < \#2[[1]], \#1, \#2]\&,$
$\{Population, (IndividualsCostValue)\}]$

$G = G + 1$

Fig. 7.33. DE outline

7.4.1 DE Flow Shop Scheduling

This section describes the application of Flow Shop scheduling as given in Fig 7.35. In this function, the obtained solution is simply passed into the *CostFunction* function.

The first variable, *JTime* accumulates the processing time of all the jobs in the first machine given as: JTime = Accumulate[#]&[Prob[[1, #]]&/@Solution];.

The second variable, *LMach*, computes the job times on all the subsequent machines iteratively. Since the maximum of the processing times is taken between the jobs:

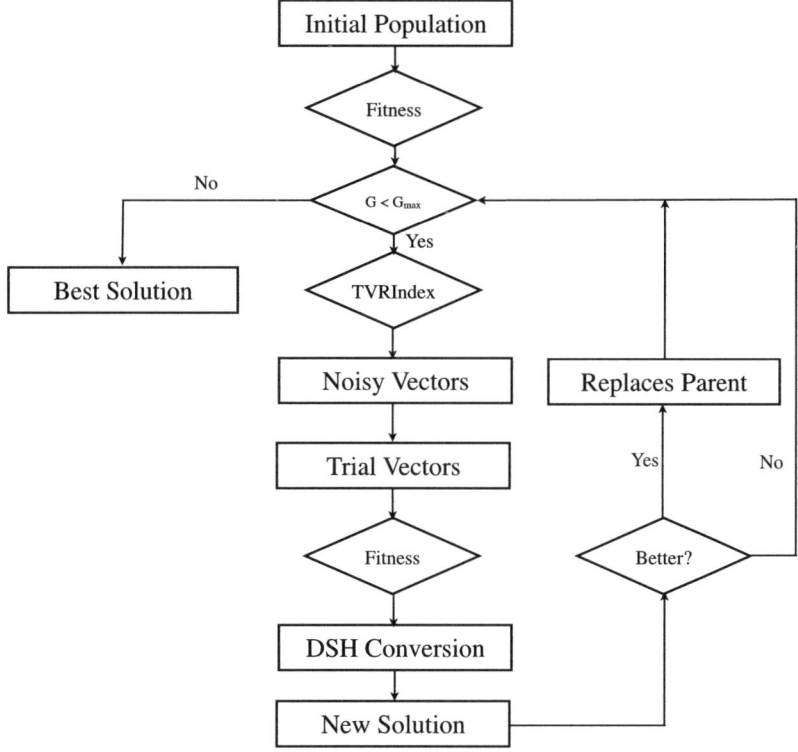

Fig. 7.34. Data flow diagram of DE

```
CostFunction = Compile[{{Solution, _Integer, 1}, {Prob, _Integer, 2}, {Mach, _Integer}},
    Module[{JTime, LMach},
     JTime = Accumulate[#] &[Prob[[1, #]] & /@ Solution];
     LMach = Accumulate[#] &[Prob[[#, Solution[[1]]]]] & /@ Range[Mach]];
         Table[JTime[[1]] = LMach[[i + 1]];
     MapIndexed[(JTime[[First[#2] + 1]] = Max[JTime[[First[#2]]], JTime[[First[#2] + 1]]] +
            Prob[[i + 1, #1]]) &, Rest[Solution]], {i, Mach - 1}]; Return[JTime[[-1]]]
    ]]
```

Fig. 7.35. Flow Shop Schedluing routine

Max[JTime[[First[#2]]], JTime[[First[#2] + 1]]] + Prob[[i + 1, #1]]) &, Rest[Solution]], the processing time value is simply accumulated in *LMach*.

For more information about Flow Shop, please see [17].

7.4.2 DE Traveling Salesman Problem

The Traveling salesman function is simply the accumulation of the distances from one city to the next. The function is given in Fig 7.36.

The first routine simply picks up the times between the cities in the *Solution*. The distance times are stored in the matrix *Distance*.

Time+ = (Distance[[Solution[[# + 1]], Solution[[#]]]]) & /@ Range[Size − 1];.

```
CostFunction = Compile[{{Solution, _Integer, 1}, {Distance, _Real, 2},
    {Size, _Integer}},
  Module[{Time = 0.0},
    Time += (Distance[[Solution[[# + 1]], Solution[[#]]]]) & /@ Range[Size - 1];
    Time += (Distance[[Solution[[1]], Solution[[Size]]]]); Return[Time]]]
```

Fig. 7.36. Traveling Salesman routine

Once all the related city distances have been added, the distance from the *last* city to the *first* city is added to complete the *tour*, given as:
Time+ = (Distance[[Solution[[1]], Solution[[Size]]]])

7.5 DE Example

The simplest approach of explaining the application of discrete set handling is to implement a worked example. In that respect, a TSP problem is proposed with only five cities, in order to make it more viable.

Assume a symmetric TSP problem given as in Table 7.2. Symmetric implies that the distances between the two cities are equal both ways of travelling.

Table 7.2. Symmetric TSP problem

Cities	A	B	C	D	E
A	0	5	10	14	24
B	5	0	5	9	19
C	10	5	0	10	14
D	14	9	10	0	10
E	24	19	14	10	0

Table 7.3. Decomposed symmetric TSP problem

Cities	A	B	C	D	E
A	**0**				
B	5	**0**			
C	10	5	**0**		
D	14	9	10	**0**	
E	24	19	14	10	**0**

Since this is a symmetric TSP problem, the *Distance Matrix* can be decomposed to the leading triangle as given in Table 7.3.

In order to use DE, some operational parameters are required, in this case the tuning parameters of *CR* and *F*, and well as the size of the population *NP* and the number of generations *Gen*. For the purpose of this example, the population is specified as 10 individuals.

7.5.1 Initialization

The first phase is the initialization of the population. Since NP has been arbitrarily set as 10, ten random permutative solutions are generated to fill the initial population as given in Table 7.4.

Table 7.4. Initial population

Solution	City 1	City 2	City 3	City 4	City 5
1	A	D	B	E	C
2	D	B	A	C	E
3	C	A	E	B	D
4	E	C	D	A	B
5	E	B	C	D	A
6	B	D	A	E	C
7	A	D	C	E	B
8	E	C	A	D	B
9	B	E	C	A	D
10	A	C	E	B	D

7.5.2 DSH Conversion

The second part is to create the discrete set for the solution. DSH assigns a raw number for each position index in the solution. In this case the most logical phase is to assign consecutive numbers for the consecutive alphabets as shown in Table 7.5.

The problem assignment now switches to the discrete set. This is given in Table 7.6.

Table 7.5. Discrete set for the cities

Cities	A	B	C	D	E
Discrete Set	1	2	3	4	5

Table 7.6. Initial Population

Solution	City 1	City 2	City 3	City 4	City 5
1	1	4	2	5	3
2	4	2	1	3	5
3	3	1	5	2	4
4	5	3	4	1	2
5	5	2	3	4	1
6	2	4	1	5	3
7	1	4	3	5	2
8	5	3	1	4	2
9	2	5	3	1	4
10	1	3	5	2	4

Table 7.7. Distance matrix for Tour 1

Cities	A	D	B	E	C
A					
D	14				
B	5	9			
E	24	9	19		
C	10	4	5	14	

Table 7.8. Fitness for the population

Solution	City 1	City 2	City 3	City 4	City 5	Fitness
1	1	4	2	5	3	66
2	4	2	1	3	5	48
3	3	1	5	2	4	48
4	5	3	4	1	2	62
5	5	2	3	4	1	72
6	2	4	1	5	3	66
7	1	4	3	5	2	62
8	5	3	1	4	2	66
9	2	5	3	1	4	66
10	1	3	5	2	4	66

7.5.3 Fitness Evaluation

The objective function for TSP is the cumulative distance between the cities, ending and starting from the same city. Taking the example of the first solution in Table 7.6; now termed Tour $1 = \{A,D,B,E,C\}$, the equivalent representation is Tour $1 = \{1,4,2,5,3\}$. The distance matrix can now be represented as in Table 7.7.

Since the tour is cyclic, the tour can further be completely represented as Tour $1 = \{A \rightarrow D \rightarrow B \rightarrow E \rightarrow C \rightarrow A\}$. From distance matrix it is now the accumulation of the tour distances Tour $1 = 14 + 9 + 19 + 14 + 10$, which gives a total of 66.

Likewise, the total tour for all the solutions is calculated and is presented in Table 7.8.

7.5.4 DE Application

The next step is the application of DE to each solution in the population. For this example the DE **Rand1Bin** strategy is selected. At this point it is important to set the DE scaling factor F. It can be given a value of 0.4.

DE application is simple. Starting from the first solution, each solution is evolved sequentially. Evolution in DE consists of a number of steps. The first step is to *randomly* select two other solutions from the population, which are unique from the solution currently under evolution. If we take the assumption that Solution 1 is currently under evolution, then we can randomly select Solution 4 and Solution 7 for example. These make the batch of *parent* solutions as given in Table 7.9.

7 Discrete Set Handling

Table 7.9. Parent solutions

Solution	City 1	City 2	City 3	City 4	City 5	Fitness
1	1	4	2	5	3	66
4	5	3	4	1	2	62
7	1	4	3	5	2	62

Table 7.10. Parent solutions crossover

Solution	City 1	City 2	City 3	City 4	City 5
1	1	4	2	5	3
4	5	3	4	1	2
7	1	4	3	5	2
Index	4	5	1	2	3

The second DE operating parameter crossover CR can now be set as 0.4. The starting point of evolution in the solution is randomly selected. In this example, solution index 3 is selected as the first variable for crossover as given in Table 7.10.

The mathematical representation of DE Rand1Bin is given as: $x_{current} + F \bullet (x_{random_1} - x_{random_2})$. x_{random_1} in this instance refers to the first randomly selected solution 4, and x_{random_2} is the second random solution 7. Since the starting index has been randomly selected as 3, the linked values for the two solutions are subtracted as 4 − 3 = 1. This value is multiplied by F, which is 0.4 The result is (1 x 0.4 = 0.4). This value is added to the current indexed solution 1: (0.4 + 2 = 2.4).

Likewise, applying the equation to the selected parent solutions yields the following values given in Table 7.11:

Table 7.11. Parent solutions final values

Solution	City 1	City 2	City 3	City 4	City 5
1	1	4	2	5	3
4	5	3	4	1	2
7	1	4	3	5	2
Index	4	5	1	2	3
Final	2.6	3.6	2.4	3.4	3

The second part shown in Table 7.12 is to select which of the new variables in the solutions will actually be accepted in the final child solution. The procedure of this is to randomly generate random numbers between 0 and 1 and if these random numbers are greater than the user specified constant CR, then these values are accepted in the child solution. Otherwise the current index values are retained.

Table 7.12. CR Application

Solution	City 1	City 2	City 3	City 4	City 5
Parent	1	4	2	5	3
Final	2.6	3.6	2.4	3.4	3
Random value	0.6	0.2	0.5	0.9	0.3

Table 7.13. Child solution

Solution	City 1	City 2	City 3	City 4	City 5
Parent	1	4	2	5	3
Child	2.6	4	2.4	3.4	3

Table 7.14. Closest Integer Approach

Solution	City 1	City 2	City 3	City 4	City 5
Child	2.6	4	2.4	3.4	3
Closest integer	3	4	2	3	3

Table 7.15. Hierarchical Approach

Solution	City 1	City 2	City 3	City 4	City 5
Child	2.6	4	2.4	3.4	3
Hierarchical Approach	2	5	1	4	3

Since *CR* has been set as 0.4, all indexes with random values greater than 0.4 are selected into the child population. The rest of the indexes are filled by the variables from the parent solution as given in Table 7.13.

Two different approaches now can be used in order to realize the *child* solution. The first is to *closest integer approach*. In this approach the integer value closest to the obtained real value is used. This is given as in Table 7.14.

The second approach is the *hierarchical approach*. In this approach, the solutions are listed according to their placement in the solution itself. This is given in Table 7.15.

The advantage of the *hierarchical approach* is that no repairment is needed to the final solution. However, it does not reflect the placements of DE values, and can be misleading. Due to this factor, the first approach of *closest integer* approach is now described.

The next step is to check if any solution exists outside of the bounds. According to [18], all out of bound variables are randomly repaired. If the case of this example all the values are within the bounds specified by the problem.

Table 7.16. Feasible solutions

Solution	City 1	City 2	City 3	City 4	City 5
Child	3	4	2	3	3
Feasible	-	**4**	**2**	-	-

Table 7.17. Final solution

Solution	City 1	City 2	City 3	City 4	City 5
Child	3	4	2	3	3
Final Solution	**3**	**4**	**2**	**1**	**5**

Table 7.18. Final solution fitness

Solution	City 1	City 2	City 3	City 4	City 5	Fitness
Final Solution	3	4	2	1	5	62

Table 7.19. DSH application

Solution	City 1	City 2	City 3	City 4	City 5
Final Solution	3	4	2	1	5
City	**C**	**D**	**B**	**A**	**E**

The final routine is to repair the solution if repetitive solutions exist. It must be stressed that not all the solutions obtained are infeasible.

The approach is to first isolate all the unique solutions as given in Table 7.16.

The missing values in this case are 1, 3, 5. Using random selection, each missing value is replaced in the final solution in Table 7.17.

Random placement is selected since it has proven highly effective [3].

The new solution is vetted for its fitness.

The new fitness of 62 improves the old fitness of the parent solution of 66 and hense the child solution is accepted in the population for the next generation. The correct arrangement is obtained by converting back using DSH into City representation as given in Table 7.19.

Using the above process, all the solutions are evolved from one generation to another. At the termination of the algorithm, the best-placed solution is retrieved.

7.6 Experimentation

All experiments have been done on the *grid cluster* of the XServers (Apple technology). Such a kind of computer technology is now commonly used for hard computing tasks.

Fig. 7.37. 1000 PC cluster 1

An example is the 1000 PCs used in genetic programming (Fig 7.37 − 7.38). In Czech Republic, there also exists such grid computers. An example of a grid configuration is the supercomputer named Amalka with 360 processors used in space research and related problems shown in Fig 7.39.

The grid cluster used for the FSS and TSP experiments, consisted of 16 XServers 2 x 2 GHz Intel Xeon, 1 GB RAM, 80 GB HD (Fig 7.40 − 7.41). Each Xserve contain 4 computational cores, so there are in total 64 computational cores. Part of the computational force has been used for FSS and TSP calculations.

7.6.1 Flow Shop Scheduling Tuning

The main issue for almost all meta-heuristics, which does optimization without knowledge of the system, is that there are parameters to tune in the algorithm. In DE, there are two control parameters, F and CR. These parameters are required in order to induce the stochastic process in the heuristic, which will enable it to find the optimal solution for that specific problem.

Fig. 7.38. 1000 PC cluster 2

[18] gave a brief outline for the different operating parameters as given in Table 7.20.

These general outlines were formulated after experimentation [18], however they were not intended for permutative problems. Since this is realized as a novel approach for DE, it becomes than imperative to create a experiment procedure for the formulation of these control values. Alongside these control values, these are altogether 7 general operating DE strategies.

1. Rand 1 Bin
2. Rand 2 Bin
3. Best 2 Bin
4. Local to Best
5. Best 1 JIter
6. Rand 1 DIter
7. Rand 1 GenDIter

Fig. 7.39. Amalka Grid

Table 7.20. Operating parameters for original DE

Control Variables	Lo	Hi	Best?	Comments
F : Scaling Factor	0	1.0+	0.3 0.9	$F \geq 0.5$
CR: Crossover probability	0	1	0.8 1.0	CR = 0, seperable
				CR = 1, epistatic

Table 7.21. Tuning Parameters

Strategy	CR	F
Rand1Bin	0.1	0.1
Rand2Bin	0.2	0.2
Best2Bin	0.3	0.3
LocaltoBest	0.4	0.4
Best1JIter	0.5	0.5
Rand1DIter	0.6	0.6
Rand1Gen DIter	0.7	0.7
	0.8	0.8
	0.9	0.9
	1	1

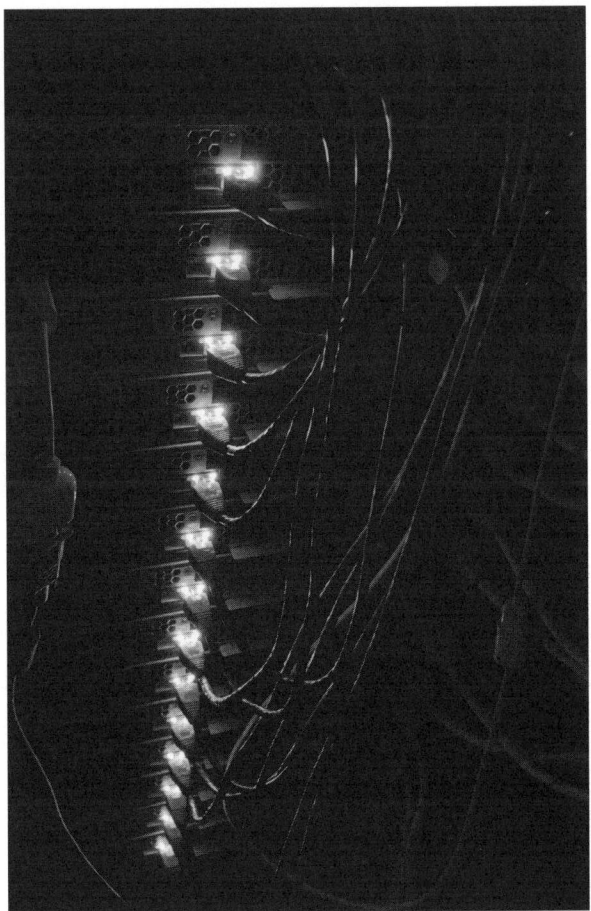

Fig. 7.40. Emanuel Cluster at UTB

Table 7.22. FSS operating parameters

Parameters	Values
Strategy	Rand 1 DIter
F	0.5
CR	0.1

So the task then is to also find the optimal operating strategy alongside the two control variables. This in itself becomes a three phase permutative problem. The sampling rate for the two control variables was kept as small as possible to 0.1.

The permutative outline for the tuning parameter is now given in Table 7.21.

Each value is permutated through the other values, so the total number of tuning experimentation conducted is 7 x 10 x 10 = 700.

Fig. 7.41. Emanuel Cluster at UTB

The second aspect is to select an appropriate test instance. For our purpose, a moderately difficult instance of 50 jobs and 20 machines from the Taillard benchmark problem set was selected.

Experimentation was conducted with Population set to 200 individuals and 100 generations allowed. The solution mesh is given in Fig 7.42.

A histogram projection in Fig 7.43. gives a better representation with the frequency of makespan.

The optimal value obtained through this experimentation is given in Table 7.22.

7.6.2 Traveling Salesman Problem Tuning

The identical tuning procedure used for Flow shop was used for parameter tuning on the Traveling Salesman Problem. Once again, 700 experimentations were conducted, and for this problem set, the moderately difficult Eil51 city problem set was selected.

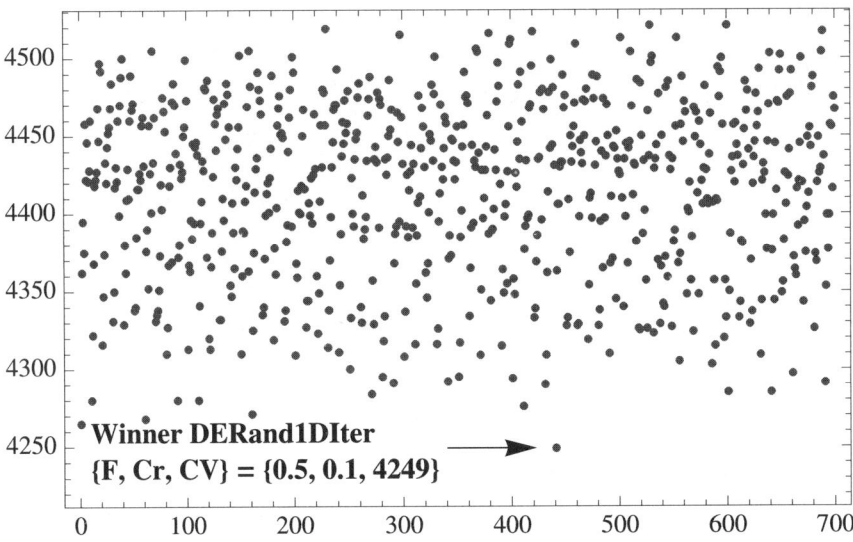

Fig. 7.42. FSS tuning graphical display

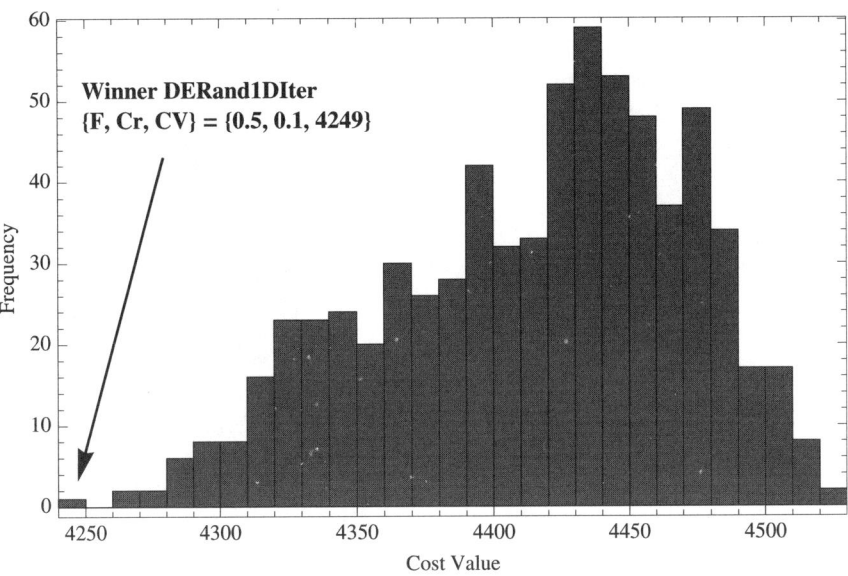

Fig. 7.43. Frequency display for FSS tuning

The solution mesh for TSP is given in Fig 7.44.
The histogram display for all the values is given in Fig 7.45.
The optimal value obtained through this experimentation is given in Table 7.23.

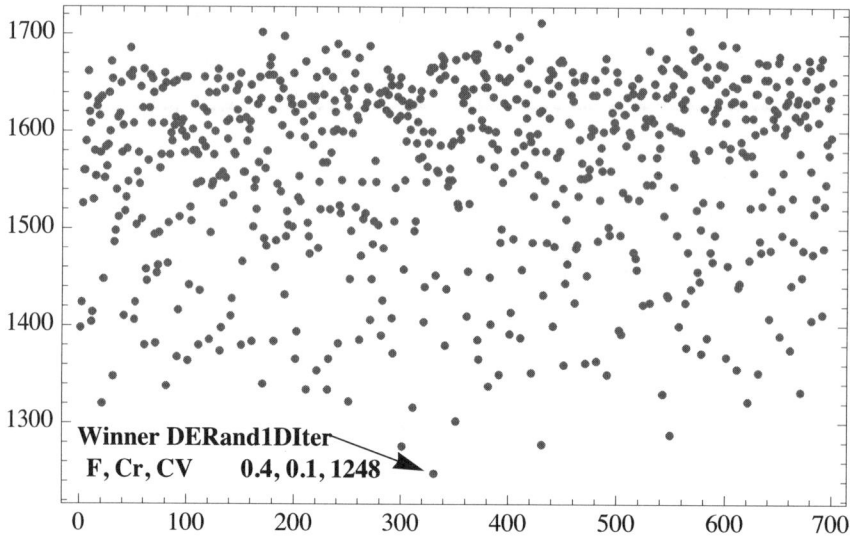

Fig. 7.44. TSP tuning graphical display

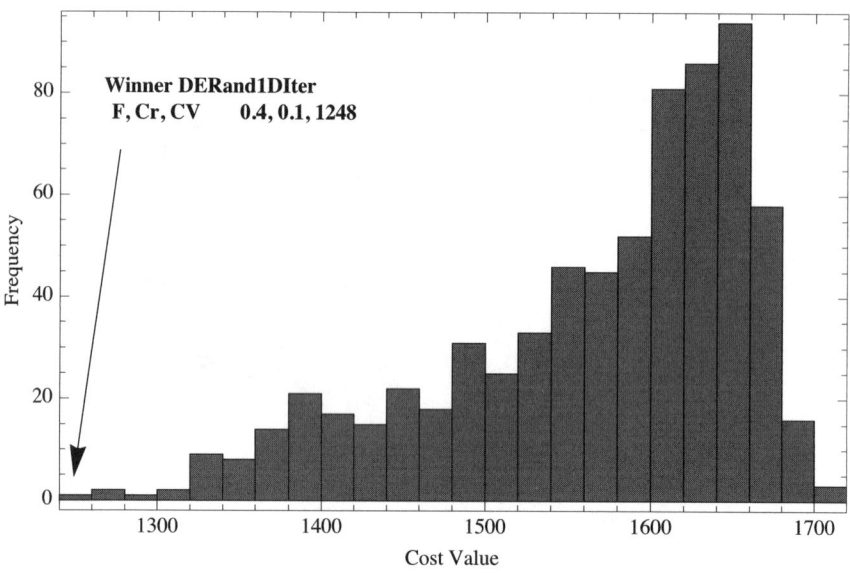

Fig. 7.45. Frequency display for TSP tuning

Using these obtained values, extensive experimentation was conducted on both the FSS and TSP problems. The core issue here is that it is shown that small changes in the control variables leads to different results. The hypothesis that parameter tuning is highly important for tuning of DE for permutative optimization is proven.

Table 7.23. FSS operating parameters

Parameters	Values
Strategy	Rand 1 Bin
F	0.4
CR	0.1

7.6.3 Flow Shop Scheduling Results

The primary experimentation was conducted on Taillard benchmark Flowshop Scheduling [21]. These sets are considered primary with a core mix of complexity and scale. Altogether 120 problem sets are involved, 10 problem instances of n job and m machine problems of 20x5, 20x10, 20x20, 50x5, 50x10, 50x20 100x5, 100x10, 100x20, 200x10, 200x20, and 500x10 are involved. For each problem instance, two bounds are given, the upper bound and the lower bound. Most reference is taken from the upper bound, which is the hypothetical optimal of a particular instance.

So the objective then is not to find the optimal solution (one can if one wants), but to gauge how effective a heuristic is over the entire range of these problems. In others words, to observe the consistency of the heuristic. To this effect, the results are presented in the following format by applying Equation 7.14.

$$\Delta_{avg} = \frac{(H-U) \bullet 100}{U} \qquad (7.14)$$

Equation 7.14 is where H represents the obtained value and U is the bound specified by [21]. The Δ_{avg}, gives the average value for all the instances in that particular class, and gives the standard deviation across all the instances. This is important in order to gauge the consistency of the heuristic.

The operating parameters of DE using Discrete Set Handling (DE_{DSH}) is given in Table 7.24. The values of CR and F were obtained through extensive parameter tuning and NP (population size) and Gen (number of generations) was kept at 700.

Table 7.24. DE_{DSH} operating parameters

Parameters	CR	F	NP	Gen
Value	0.5	0.1	500	700

The collated results are presented in Table 7.25. These results are presented with the results compiled by [22].

Generally, two classes of heuristics are observed: those, which are canonical, and those, which have embedded local search. To the first class of heuristics belong GA (Genetic Algorithm), PSO_{spv} (Particle Swamp Optimization with smallest position value) and DE_{spv} (Differential Evolution with smallest position value). The second class has $DE_{spv+exchange}$, which is DE_{spv} with local search.

Table 7.25. Flowshop scheduling results

	GA		PSO_{spv}		DE_{spv}		$DE_{spv+exchange}$		DE_{DSH}		$DE_{DSH+EXH}$	
	$\|\Delta_{avg}$	Δ_{std}	$\|\Delta_{avg}$	Δ_{std}	$\|\Delta_{avg}$	Δ_{std}	$\|\Delta_{avg}$	Δ_{std}	$\|\Delta_{avg}$	Δ_{std}	$\|\Delta_{avg}$	Δ_{std}
20x5	3.13	1.86	1.71	1.25	2.25	1.37	0.69	0.64	1.2	0.42	1.07	0.55
20x10	5.42	1.72	3.28	1.19	3.71	1.24	2.01	0.93	2.5	0.41	2.35	0.6
20x20	4.22	1.31	2.84	1.15	3.03	0.98	1.85	0.87	2.52	0.32	1.92	0.53
50x5	1.69	0.79	1.15	0.7	0.88	0.52	0.41	0.37	0.84	0.56	0.5	0.56
50x10	5.61	1.41	4.83	1.16	4.12	1.1	2.41	0.9	5.09	1.02	3.21	1.11
50x20	6.95	1.09	6.68	1.35	5.56	1.22	3.59	0.78	7.05	1.08	4.21	0.85
100x5	0.81	0.39	0.59	0.34	0.44	0.29	0.21	0.21	0.73	0.32	0.32	0.24
100x10	3.12	0.95	3.26	1.04	2.28	0.75	1.41	0.57	3.11	1.2	1.5	1.08
100x20	6.32	0.89	7.19	0.99	6.78	1.12	3.11	0.55	5.98	0.57	4.19	0.82
200x10	2.08	0.45	2.47	0.71	1.88	0.69	1.06	0.35	3.77	1.31	1.781	1.1
200x20									9.82	0.7	4.32	0.68
500x10									6.28	0.39	4.13	0.41

Table 7.26. Comparison results of heuristics without local search

	GA		PSO_{spv}		DE_{spv}		DE_{DSH}	
	$\|\Delta_{avg}$	Δ_{std}	$\|\Delta_{avg}$	Δ_{std}	$\|\Delta_{avg}$	Δ_{std}	$\|\Delta_{avg}$	Δ_{std}
20x5	3.13	1.86	1.71	1.25	2.25	1.37	**1.2**	0.42
20x10	5.42	1.72	3.28	1.19	3.71	1.24	**2.5**	0.41
20x20	4.22	1.31	2.84	1.15	3.03	0.98	**2.52**	0.32
50x5	1.69	0.79	1.15	0.7	0.88	0.52	**0.84**	0.56
50x10	5.61	1.41	4.83	1.16	**4.12**	1.1	5.09	1.02
50x20	6.95	1.09	6.68	1.35	**5.56**	1.22	7.05	1.08
100x5	0.81	0.39	0.59	0.34	**0.44**	0.29	0.73	0.32
100x10	3.12	0.95	3.26	1.04	**2.28**	0.75	3.11	1.2
100x20	6.32	0.89	7.19	0.99	6.78	1.12	**5.98**	0.57
200x10	2.08	0.45	2.47	0.71	1.88	0.69	3.77	1.31
200x20							9.82	0.7
500x10							6.28	0.39

The experimentation of $DE_{DSH+EXH}$ was done on two parts, one with local search and one without. The comparison result of DE_{DSH} is given in Table 7.26.

DE_{DSH} was able to find the better average values for the problem sets of 20x5, 20x10, 20x20, 50x5 and 100x20. The others sets was dominated by DE_{spv}. A graphical output for the different sets is given in Fig 7.46. The deviation output is given in Fig 7.47.

The second set is the comparison of the heuristics with local search, namely $DE_{spv+exchange}$ and $DE_{DSH+EXC}$ as presented in Table 7.27.

As observed $DE_{spv+exchange}$ is the better performing heuristic. The last two columns gives the analysis comparisons and on average $DE_{DSH+EXH}$ is only 0.42% away from $DE_{spv+exchange}$. The graphical displays are given in Figs 7.48 and 7.49.

7 Discrete Set Handling   199

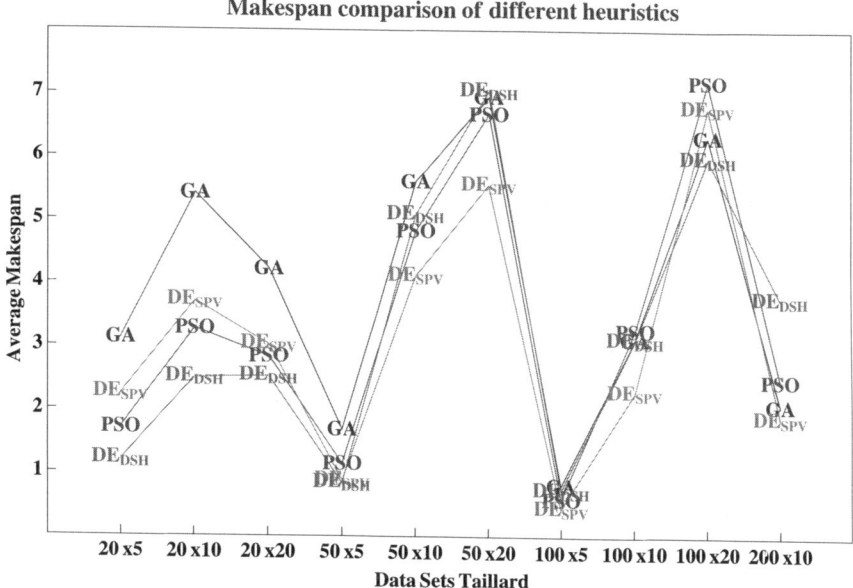

Fig. 7.46. Makespan display of different heuristics without local search

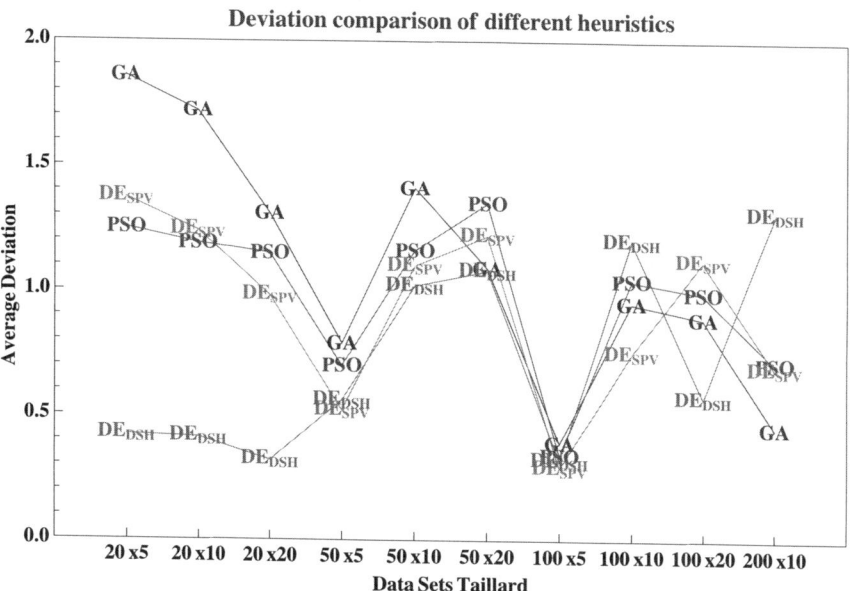

Fig. 7.47. Deviation display of different heuristics without local search

Table 7.27. Comparison results of heuristics with local search

	$DE_{spv+exchange}$		$DE_{DSH+EXH}$		Analysis	
	Δ_{avg}	Δ_{std}	Δ_{avg}	Δ_{std}	Δ_{avg}	Δ_{std}
20x5	0.69	0.64	1.07	0.42	0.38	0.22
20x10	2.01	0.93	2.35	0.41	0.34	0.52
20x20	1.85	0.87	1.92	0.32	0.06	0.55
50x5	0.41	0.37	0.5	0.56	0.09	0.19
50x10	2.41	0.9	3.21	1.02	0.8	0.12
50x20	3.59	0.78	4.21	1.08	0.62	0.3
100x5	0.21	0.21	0.32	0.32	0.11	0.11
100x10	1.41	0.57	1.5	1.2	0.09	0.63
100x20	3.11	0.55	4.19	0.57	1.08	0.02
200x10	1.06	0.35	1.78	1.31	0.72	0.96
200x20			4.32	0.7		
500x10			4.13	0.39		
Average					0.42	0.361

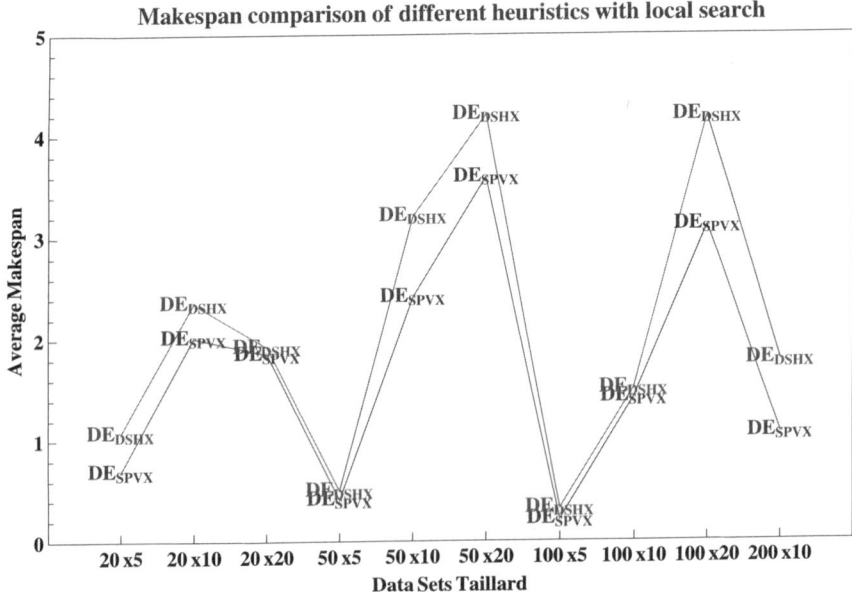

Fig. 7.48. Makespan display of different heuristics with local search

In terms of average deviation, $DE_{DSH+EXH}$ generally has better values than $DE_{spv+exchange}$. This implies that $DE_{DSH+EXC}$ obtains solutions with greater regularity and consistency than $DE_{spv+exchange}$.

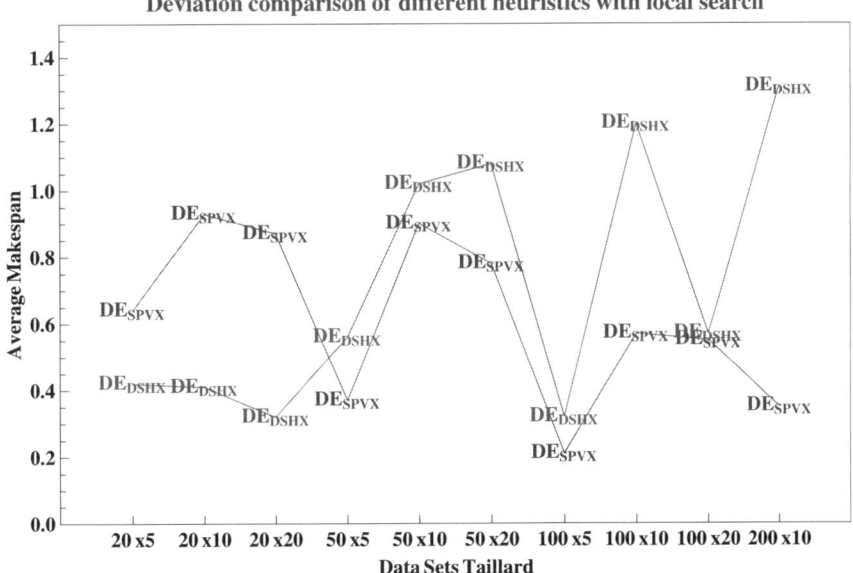

Fig. 7.49. Deviation display of different heuristics with local search

7.6.4 Traveling Salesman Problem Results

7.6.4.1 Symmetric Traveling Salesman

The second set of problem set to be considered is the Traveling Salesman Problem (TSP). TSP is a widely realized problem with many applications in real life problems. However, to compensate for its myriad usage, a number of targeted heuristics have evolved to solve it; often to optimal as is the case for all for all known problem instance in the TSPLIB. For evolutionary heuristics to operate in TSP, it has become a norm for them to employ local search, usually 3 opt [4]. Utilizing local search heuristics always improve the quality of the results of the solutions, since triangle inequality rule and Lin−Kernigham are very robust *deterministic* search heuristics.

The operating parameters for the TSP is given in Table 7.28.

A sample of TSP problem is given in Table 7.28. Comparison is done with the Ant Colony (AC), Simulated Annealing (SA), Self Organising Map (SOM) and Furthest Insertion (FI) of [4].

$$\Delta_{avg} = \frac{H - U}{U} \quad (7.15)$$

Table 7.28. DE_{DSH} TSP operating parameters

Parameters	CR	F	NP	Gen
Value	0.4	0.1	500	700

Table 7.29. STSP comparison results

Instance	Optimal	ACS_{3opt}	SA_{3opt}	SOM	FI_{3opt}	DE_{DSH}
City 1	5.84	0	0	0	0.002	0.002
City 2	5.99	0.002	0	0.002	0	0.03
City 3	5.57	0	0	0.002	0	0.059
City 4	5.06	0.12	0.12	0	0.13	0.16
City 5	6.17	0	0	0.003	0.03	0.01

Table 7.30. General STSP comparison results

Instance	Optimal	ACS_{3opt}	DE_{DSH}
att532	27,686	0	0.17
d198	15,780	0.006	0.54
eil51	426	-	0.08
eil76	538	-	0.1
fl1577	8,806	0.03	1.23
kroA100	21,282	0	0.56
pcb442	50,779	0.01	0.32
rat783	8,806	-	0.92
Average			**0.49**

The results are presented in Table 7.29 as percentage increase upon the reported optimal as given in Equation 7.15.

In this instance, DE_{DSH}, was competitive to the other performing heuristics. As shown, no one heuristic was able to find all optimal values, and some heuristic performed better than other for specific instances.

The second set of experiment was conducted on some selective TSP instances [23]. The results are presented in Table 7.30.

The comparison is done with ACS_{3opt} of [4]. ACS performs very well, almost achieving the optimal solution. DE_{DSH} performs well, obtaining on average 0.49% to the optimal results for the entire set. The set contains instance's ranging from sizes of 51 to 1577 cities.

7.6.4.2 Asymmetric Traveling Salesman

The second set of problems is that, which involves the asymmetric TSP. Asymmetric TSP is one where the distances between two cities are not equal, to and from. This implies that going from one city to another has a different distance than coming back from that city to the original one. The results are presented in Table 7.31.

The results for ATSP are on average 1.112% over the optimal value. However, it should be noted that the experimentation values was kept stagnant to fixed values, even as the problem size was increased, hence the trend of worsening solutions as problem size increases.

Table 7.31. General ATSP comparison results

Instance	Optimal	ACS_{3opt}	DE_{DSH}
ft70	38673	0.001	0.96
ftv170	2755	0.002	2.32
kro124p	36230	0	1.57
p43	5620	0	0.24
ry48p	14422	0	0.47
Average			**1.112**

7.7 Conclusion

Differential Evolution is an effective heuristic for optimization. This approach was an attempt to show it effectiveness in permutative problems. The key approach was to keep the conversion of the operational domain as simple as possible, as shown in this variant of discrete set handling. Simplicity removes excess computation overhead to this heuristic while at the same time delivering comparative results.

Two different problem scopes of Flow Shop scheduling and Traveling Salesman problems were attempted. This was done in order to show that this generic version of DE is able to work in different classes of problems, and not simply tailor made for a special class. The core research focused on Flow Shop with Traveling Salesman providing a secondary comparison.

A principle direction as seen in this research has been the tuning of the heuristic. Researchers, who generally take the default values, often overlook this process, however it is imperative to check for the best values. As shown from the obtained results, the operating values obtained for the two different problems were unique in all aspects.

The results obtained can be visualized as competitive for their own classes. The most promising is the results obtained for Flow Shop, and the worst performing is the Asymmetric Traveling Salesman. It is believed that a better local search heuristic, like Lin–Kernighan or a 3 Opt heuristic will further improve the quality of the solutions.

Further directions for this approach will involve further testing with other problem classes like Vehicle Routing and Quadratic Assignment, which are also realised in real systems.

Acknowledgement

This work was supported by grant No. MSM 7088352101 of the Ministry of Education of the Czech Republic and by grants of the Grant Agency of the Czech Republic GACR 102/06/1132.

References

1. Bland, G., Shallcross, D.: Large traveling salesman problems arising fromexperiments in X-ray crystallography: A preliminary report on computation. OpersRes. Lett. 8, 125–128 (1989)
2. Croes, G.: A method for solving traveling salesman problems. Oper. Res. 6, 791–812 (1958)

3. Davendra, D., Onwubolu, G.: Enhanced Differential Evolution hybrid Scatter Search for Discrete Optimisation. In: Proceeding of the IEEE Congress on Evolutionary Computation, Singapore, September 25-28, pp. 1156–1162 (2007)
4. Dorigo, M., Gambardella, M.: Ant Colony System: A Cooperative Learning Approach to the Traveling Salesman Problem. IEEE Trans. Evol. Comput. 1(1), 53–66 (1997)
5. Flood, M.: The traveling-salesman problem. Oper. Res. 4, 61–75 (1956)
6. Johnson, G.: Artificial immune systems programming for symbolic regression. In: Ryan, C., Soule, T., Keijzer, M., Tsang, E.P.K., Poli, R., Costa, E. (eds.) EuroGP 2003. LNCS, vol. 2610, pp. 345–353. Springer, Heidelberg (2003)
7. Korte, B.: Applications of combinatorial optimization. In: The 13th International Mathematical Programming Symposium, Tokyo (1988)
8. Koza, J.: Genetic Programming: A paradigm for genetically breeding populations of computer programs to solve problems. Stanford University, Computer Science Department, Technical Report STAN–CS–90–1314 (1990)
9. Koza, J.: Genetic Programming. MIT Press, Boston (1998)
10. Koza, J., Bennet, F., Andre, D., Keane, M.: Genetic Programming III. Morgan Kaufnamm, New York (1999)
11. Koza, J., Keane, M., Streeter, M.: Evolving inventions. Sci. Am., 40–47 (February 2003)
12. Garey, M., Johnson, D.: Computers and Intractability: A Guide to the Theory of NP-Completeness. W. H. Freeman, San Francisco (1979)
13. O/Neill, M., Ryan, C.: Grammatical Evolution. In: Evolutionary Automatic Programming in an Arbitrary Language. Springer, New York (2003)
14. O/Neill, M., Brabazon, A.: Grammatical Differential Evolution. In: Proc. International Conference on Artificial Intelligence (ICAI 2006), pp. 231–236. CSEA Press (2006)
15. Oplatkova, Z., Zelinka, I.: Investigation on artificial ant using analytic programming. In: Proc. Genetic and Evolutionary Computation Conference 2006, Seattle, WA, pp. 949–950 (2006)
16. Operations Reserach Library (Cited September 1, 2008), http://people.brunel.ac.uk/~mastjjb/jeb/info.htm
17. Pinedo, M.: Scheduling: theory, algorithms and systems. Prentice Hall, Inc., New Jersey (1995)
18. Price, K.: An introduction to differential evolution. In: Corne, D., Dorigo, M., Glover, F. (eds.) New Ideas in Optimisation, pp. 79–108. McGraw Hill, International (1999)
19. Ryan, C., Collins, J., O'Neill, M.: Grammatical evolution: Evolving programs for an arbitrary language. In: Banzhaf, W., Poli, R., Schoenauer, M., Fogarty, T.C. (eds.) EuroGP 1998. LNCS, vol. 1391, p. 83. Springer, Heidelberg (1998)
20. Lin, S., Kernighan, B.: An Effective Heuristic Algorithm for the Traveling-Salesman Problem. Oper. Res. 21, 498–516 (1973)
21. Taillard, E.: Benchmarks for basic scheduling problems. Eur. J. Oper. Res. 64, 278–285 (1993)
22. Tasgetiren, M., Liang, Y.-C., Sevkli, M., Gencyilmaz, G.: Differential Evolution Algorithm for Permutative Flowshops Sequencing Problem with Makespan Criterion. In: 4th International Symposium on Intelligent Manufacturing Systems. IMS 2004, Sakaraya, Turkey, September 5–8, pp. 442–452 (2004)
23. TSPLIB (Cited September 1, 2008), http://elib.zib.de/pub/mp-testdata/tsp/tsplib/tsplib.html
24. Zelinka, I.: Analytic programming by Means of new evolutionary algorithms. In: Proc. 1st International Conference on New Trends in Physics 2001, Brno, Czech Republic, pp. 210–214 (2001)

25. Zelinka, I.: Analytic programming by means of soma algorithm. In: ICICIS 2002, First International Conference on Intelligent Computing and Information Systems, Cairo, Egypt, pp. 148–154 (2002)
26. Zelinka, I., Oplatkova, Z.: Analytic programming – Comparative study. In: Proc. the Second International Conference on Computational Intelligence, Robotics, and Autonomous Systems, Singapore, paper No. PS04-2-04 (2003)
27. Zelinka, I., Oplatkova, Z., Nolle, L.: Analytic programming – Symbolic regression by means of arbitrary evolutionary algorithms. Int. J. Simulat. Syst. Sci. Tech. 6(9), 44–56 (2005)
28. Zelinka, I., Chen, G., Celikovsky, S.: Chaos sythesis by menas of evolutionary algorithms. Int. J. Bifurcat Chaos 4, 911–942 (2008)

A
Smallest Position Value Approach

A.1 Clusters for the Instance 11EIL51

Table A.1. Clusters for the Instance 11EIL51

Cluster	Node						
N_1	19	40	41				
N_2	3	20	35	36			
N_3	24	43					
N_4	33	39					
N_5	11	12	27	32	46	47	51
N_6	2	16	21	29	34	50	
N_7	8	22	26	28	31		
N_8	13	14	18	25			
N_9	4	15	17	37	42	44	45
N_{10}	1	6	7	23	48		
N_{11}	5	9	10	30	38	49	

A.2 Pseudo Code for Distance Calculation

Double x[MaxNode], y[MaxNode], xd, yd, $distance$[MaxNode][MaxNode];
For(i = 0; i < **MaxNode**; i++)**Do**
  For(j = i; j < **MaxNode**; j++)**Do**
    $xd = x[i] - x[j]$;
    $yd = y[i] - y[j]$;
    $distance[j][i] = distance[i][j] =$ **nint**(**sqrt**($xd * xd + yd * yd$));
  EndDo
EndDo

nint: round to the nearest integer.

A.3 Distance (ij, d_{ij}) Information for the Instance 11EIL51

1 1 0	1 2 12	1 3 19	1 4 31	1 5 22	1 6 17	1 7 23	1 8 12	1 9 24	1 10 34	1 11 12	1 12 21
1 13 42	1 14 27	1 15 36	1 16 19	1 17 31	1 18 28	1 19 46	1 20 21	1 21 27	1 22 7	1 23 22	1 24 29
1 25 33	1 26 19	1 27 8	1 28 16	1 29 21	1 30 33	1 31 17	1 32 6	1 33 43	1 34 31	1 35 27	1 36 31
1 37 30	1 38 19	1 39 43	1 40 56	1 41 44	1 42 45	1 43 34	1 44 38	1 45 42	1 46 14	1 47 23	1 48 12
1 49 26	1 50 24	1 51 14	2 1 12	2 2 0	2 3 15	2 4 37	2 5 21	2 6 28	2 7 35	2 8 22	2 9 16
2 10 28	2 11 11	2 12 25	2 13 50	2 14 38	2 15 35	2 16 9	2 17 34	2 18 36	2 19 51	2 20 12	2 21 15
2 22 11	2 23 34	2 24 41	2 25 43	2 26 29	2 27 19	2 28 19	2 29 9	2 30 24	2 31 23	2 32 11	2 33 39
2 34 20	2 35 19	2 36 24	2 37 32	2 38 15	2 39 35	2 40 62	2 41 50	2 42 48	2 43 46	2 44 39	2 45 40
2 46 20	2 47 29	2 48 25	2 49 21	2 50 14	2 51 21	3 1 19	3 2 15	3 3 0	3 4 50	3 5 36	3 6 35
3 7 35	3 8 21	3 9 31	3 10 43	3 11 25	3 12 38	3 13 61	3 14 46	3 15 51	3 16 23	3 17 48	3 18 47
3 19 64	3 20 8	3 21 24	3 22 12	3 23 37	3 24 46	3 25 52	3 26 25	3 27 27	3 28 9	3 29 17	3 30 37
3 31 16	3 32 23	3 33 54	3 34 32	3 35 10	3 36 12	3 37 47	3 38 30	3 39 49	3 40 75	3 41 63	3 42 62
3 43 47	3 44 54	3 45 56	3 46 32	3 47 42	3 48 28	3 49 36	3 50 27	3 51 33	4 1 31	4 2 37	4 3 50
4 4 0	4 5 20	4 6 21	4 7 37	4 8 38	4 9 33	4 10 31	4 11 27	4 12 13	4 13 15	4 14 18	4 15 19
4 16 35	4 17 8	4 18 8	4 19 15	4 20 49	4 21 45	4 22 38	4 23 31	4 24 29	4 25 18	4 26 43	4 27 24
4 28 47	4 29 44	4 30 38	4 31 46	4 32 27	4 33 31	4 34 42	4 35 56	4 36 61	4 37 13	4 38 27	4 39 41
4 40 25	4 41 13	4 42 16	4 43 41	4 44 15	4 45 25	4 46 18	4 47 8	4 48 29	4 49 28	4 50 38	4 51 17
5 1 22	5 2 21	5 3 36	5 4 20	5 5 0	5 6 25	5 7 40	5 8 33	5 9 12	5 10 14	5 11 11	5 12 9
5 13 35	5 14 30	5 15 15	5 16 16	5 17 15	5 18 23	5 19 32	5 20 33	5 21 25	5 22 27	5 23 36	5 24 39
5 25 34	5 26 40	5 27 21	5 28 37	5 29 25	5 30 18	5 31 39	5 32 16	5 33 21	5 34 21	5 35 40	5 36 45
5 37 11	5 38 7	5 39 24	5 40 42	5 41 33	5 42 28	5 43 49	5 44 18	5 45 20	5 46 12	5 47 15	5 48 29
5 49 8	5 50 17	5 51 14	6 1 17	6 2 28	6 3 35	6 4 21	6 5 25	6 6 0	6 7 16	6 8 18	6 9 34
6 10 40	6 11 22	6 12 18	6 13 27	6 14 10	6 15 34	6 16 32	6 17 25	6 18 15	6 19 35	6 20 38	6 21 41
6 22 23	6 23 11	6 24 14	6 25 17	6 26 22	6 27 9	6 28 30	6 29 37	6 30 42	6 31 27	6 32 17	6 33 45
6 34 42	6 35 44	6 36 47	6 37 27	6 38 27	6 39 50	6 40 44	6 41 32	6 42 37	6 43 23	6 44 33	6 45 41
6 46 14	6 47 16	6 48 9	6 49 33	6 50 36	6 51 11	7 1 23	7 2 35	7 3 35	7 4 37	7 5 40	7 6 16
7 7 0	7 8 14	7 9 46	7 10 54	7 11 33	7 12 34	7 13 40	7 14 22	7 15 51	7 16 41	7 17 41	7 18 30
7 19 50	7 20 40	7 21 50	7 22 26	7 23 6	7 24 14	7 25 27	7 26 11	7 27 20	7 28 26	7 29 44	7 30 55
7 31 21	7 32 27	7 33 60	7 34 53	7 35 45	7 36 46	7 37 44	7 38 40	7 39 64	7 40 58	7 41 47	7 42 53
7 43 12	7 44 50	7 45 57	7 46 28	7 47 32	7 48 11	7 49 47	7 50 47	7 51 26	8 1 12	8 2 22	8 3 21
8 4 38	8 5 33	8 6 18	8 7 14	8 8 0	8 9 36	8 10 46	8 11 24	8 12 30	8 13 45	8 14 28	8 15 46
8 16 30	8 17 39	8 18 32	8 19 52	8 20 26	8 21 37	8 22 12	8 23 16	8 24 25	8 25 34	8 26 7	8 27 14
8 28 13	8 29 30	8 30 44	8 31 9	8 32 17	8 33 54	8 34 42	8 35 31	8 36 33	8 37 40	8 38 30	8 39 55
8 40 62	8 41 50	8 42 53	8 43 26	8 44 47	8 45 53	8 46 23	8 47 31	8 48 9	8 49 38	8 50 35	8 51 22
9 1 24	9 2 16	9 3 31	9 4 33	9 5 12	9 6 34	9 7 46	9 8 36	9 9 0	9 10 12	9 11 13	9 12 21
9 13 48	9 14 41	9 15 23	9 16 8	9 17 27	9 18 35	9 19 44	9 20 25	9 21 13	9 22 26	9 23 43	9 24 48
9 25 45	9 26 43	9 27 27	9 28 35	9 29 16	9 30 8	9 31 39	9 32 19	9 33 24	9 34 9	9 35 32	9 36 38
9 37 23	9 38 7	9 39 19	9 40 54	9 41 45	9 42 39	9 43 56	9 44 28	9 45 26	9 46 21	9 47 27	9 48 35
9 49 6	9 50 6	9 51 23	10 1 34	10 2 28	10 3 43	10 4 31	10 5 14	10 6 40	10 7 54	10 8 46	10 9 12
10 10 0	10 11 22	10 12 23	10 13 46	10 14 44	10 15 16	10 16 20	10 17 24	10 18 36	10 19 39	10 20 37	10 21 24
10 22 37	10 23 50	10 24 53	10 25 47	10 26 53	10 27 34	10 28 47	10 29 28	10 30 9	10 31 50	10 32 28	10 33 12
10 34 16	10 35 43	10 36 44	10 37 19	10 38 15	10 39 10	10 40 48	10 41 41	10 42 32	10 43 63	10 44 22	10 45 16
10 46 26	10 47 28	10 48 43	10 49 8	10 50 17	10 51 28	11 1 12	11 2 11	11 3 25	11 4 27	11 5 11	11 6 22
11 7 33	11 8 24	11 9 13	11 10 22	11 11 0	11 12 14	11 13 40	11 14 30	11 15 26	11 16 10	11 17 23	11 18 26
11 19 40	11 20 23	11 21 20	11 22 16	11 23 31	11 24 36	11 25 35	11 26 31	11 27 14	11 28 26	11 29 17	11 30 21
11 31 28	11 32 14	11 33 31	11 34 21	11 35 30	11 36 35	11 37 21	11 38 7	11 39 31	11 40 51	11 41 40	11 42 37
11 43 44	11 44 29	11 45 31	11 46 10	11 47 19	11 48 22	11 49 14	11 50 15	11 51 12	12 1 21	12 2 25	12 3 38
12 4 13	12 5 9	12 6 18	12 7 34	12 8 30	12 9 21	12 10 23	12 11 14	12 12 0	12 13 27	12 14 21	12 15 17
12 16 23	12 17 10	12 18 14	12 19 26	12 20 37	12 21 33	12 22 27	12 23 29	12 24 30	12 25 25	12 26 36	12 27 16
12 28 37	12 29 31	12 30 27	12 31 37	12 32 16	12 33 27	12 34 30	12 35 44	12 36 49	12 37 10	12 38 14	12 39 33
12 40 37	12 41 26	12 42 24	12 43 41	12 44 17	12 45 23	12 46 7	12 47 6	12 48 24	12 49 17	12 50 25	12 51 8
13 1 42	13 2 50	13 3 61	13 4 15	13 5 35	13 6 27	13 7 40	13 8 45	13 9 48	13 10 46	13 11 40	13 12 27
13 13 0	13 14 18	13 15 32	13 16 50	13 17 22	13 18 14	13 19 14	13 20 62	13 21 59	13 22 49	13 23 34	13 24 27
13 25 13	13 26 48	13 27 34	13 28 57	13 29 58	13 30 53	13 31 54	13 32 39	13 33 44	13 34 57	13 35 69	13 36 73
13 37 27	13 38 41	13 39 55	13 40 19	13 41 9	13 42 22	13 43 39	13 44 27	13 45 37	13 46 30	13 47 21	13 48 36
13 49 43	13 50 52	13 51 29	14 1 27	14 2 38	14 3 46	14 4 18	14 5 30	14 6 10	14 7 22	14 8 28	14 9 41
14 10 44	14 11 30	14 12 21	14 13 18	14 14 0	14 15 35	14 16 40	14 17 24	14 18 10	14 19 29	14 20 48	14 21 50
14 22 34	14 23 16	14 24 11	14 25 6	14 26 30	14 27 19	14 28 40	14 29 46	14 30 48	14 31 37	14 32 26	14 33 47
14 34 50	14 35 54	14 36 58	14 37 28	14 38 34	14 39 54	14 40 37	14 41 25	14 42 33	14 43 23	14 44 32	14 45 42
14 46 20	14 47 16	14 48 18	14 49 39	14 50 44	14 51 18	15 1 36	15 2 35	15 3 51	15 4 19	15 5 15	15 6 34
15 7 51	15 8 46	15 9 23	15 10 16	15 11 26	15 12 17	15 13 32	15 14 35	15 15 0	15 16 30	15 17 11	15 18 25

Smallest Position Value Approach

15 19 23	15 20 47	15 21 37	15 22 41	15 23 46	15 24 46	15 25 36	15 26 53	15 27 33	15 28 51	15 29 39	15 30 25
15 31 53	15 32 30	15 33 12	15 34 30	15 35 54	15 36 59	15 37 7	15 38 21	15 39 23	15 40 33	15 41 26	15 42 16
15 43 57	15 44 6	15 45 7	15 46 23	15 47 19	15 48 41	15 49 17	15 50 29	15 51 25	16 1 19	16 2 9	16 3 23
16 4 35	16 5 16	16 6 32	16 7 41	16 8 30	16 9 8	16 10 20	16 11 10	16 12 23	16 13 50	16 14 40	16 15 30
16 16 0	16 17 31	16 18 36	16 19 48	16 20 18	16 21 10	16 22 19	16 23 39	16 24 45	16 25 45	16 26 37	16 27 23
16 28 28	16 29 9	16 30 15	16 31 32	16 32 15	16 33 32	16 34 12	16 35 24	16 36 30	16 37 28	16 38 9	16 39 27
16 40 59	16 41 48	16 42 44	16 43 52	16 44 34	16 45 34	16 46 20	16 47 28	16 48 30	16 49 14	16 50 6	16 51 22
17 1 31	17 2 34	17 3 48	17 4 8	17 5 15	17 6 25	17 7 41	17 8 39	17 9 27	17 10 24	17 11 23	17 12 10
17 13 22	17 14 24	17 15 11	17 16 31	17 17 0	17 18 14	17 19 17	17 20 46	17 21 40	17 22 37	17 23 36	17 24 35
17 25 25	17 26 45	17 27 25	17 28 47	17 29 40	17 30 31	17 31 47	17 32 25	17 33 23	17 34 35	17 35 53	17 36 58
17 37 5	17 38 22	17 39 33	17 40 28	17 41 18	17 42 14	17 43 47	17 44 9	17 45 18	17 46 17	17 47 9	17 48 32
17 49 22	17 50 32	17 51 17	18 1 28	18 2 36	18 3 47	18 4 8	18 5 23	18 6 15	18 7 30	18 8 32	18 9 35
18 10 36	18 11 26	18 12 14	18 13 14	18 14 10	18 15 25	18 16 36	18 17 14	18 18 0	18 19 20	18 20 47	18 21 46
18 22 35	18 23 24	18 24 21	18 25 11	18 26 36	18 27 20	18 28 43	18 29 44	18 30 41	18 31 41	18 32 25	18 33 37
18 34 44	18 35 54	18 36 58	18 37 19	18 38 28	18 39 46	18 40 30	18 41 17	18 42 23	18 43 33	18 44 22	18 45 32
18 46 16	18 47 8	18 48 23	18 49 31	18 50 39	18 51 15	19 1 46	19 2 51	19 3 64	19 4 15	19 5 32	19 6 35
19 7 50	19 8 52	19 9 44	19 10 39	19 11 40	19 12 26	19 13 14	19 14 29	19 15 23	19 16 48	19 17 17	19 18 20
19 19 0	19 20 63	19 21 57	19 22 53	19 23 44	19 24 39	19 25 26	19 26 57	19 27 39	19 28 62	19 29 57	19 30 47
19 31 61	19 32 41	19 33 33	19 34 52	19 35 70	19 36 75	19 37 21	19 38 39	19 39 46	19 40 11	19 41 5	19 42 9
19 43 52	19 44 17	19 45 26	19 46 32	19 47 22	19 48 44	19 49 38	19 50 49	19 51 32	20 1 21	20 2 12	20 3 8
20 4 49	20 5 33	20 6 38	20 7 40	20 8 26	20 9 25	20 10 37	20 11 23	20 12 37	20 13 62	20 14 48	20 15 47
20 16 18	20 17 46	20 18 47	20 19 63	20 20 0	20 21 17	20 22 15	20 23 41	20 24 49	20 25 54	20 26 32	20 27 29
20 28 17	20 29 10	20 30 31	20 31 23	20 32 22	20 33 49	20 34 25	20 35 7	20 36 13	20 37 44	20 38 26	20 39 43
20 40 74	20 41 62	20 42 60	20 43 52	20 44 51	20 45 51	20 46 31	20 47 41	20 48 32	20 49 31	20 50 21	20 51 32
21 1 27	21 2 15	21 3 24	21 4 45	21 5 25	21 6 41	21 7 50	21 8 37	21 9 13	21 10 24	21 11 20	21 12 33
21 13 59	21 14 50	21 15 37	21 16 10	21 17 40	21 18 46	21 19 57	21 20 17	21 21 0	21 22 25	21 23 48	21 24 55
21 25 55	21 26 44	21 27 33	21 28 31	21 29 7	21 30 16	21 31 37	21 32 24	21 33 36	21 34 9	21 35 21	21 36 27
21 37 36	21 38 18	21 39 27	21 40 67	21 41 58	21 42 52	21 43 61	21 44 42	21 45 39	21 46 30	21 47 38	21 48 39
21 49 20	21 50 8	21 51 32	22 1 7	22 2 11	22 3 12	22 4 38	22 5 27	22 6 23	22 7 26	22 8 12	22 9 26
22 10 37	22 11 16	22 12 27	22 13 49	22 14 34	22 15 41	22 16 19	22 17 37	22 18 35	22 19 53	22 20 15	22 21 25
22 22 0	22 23 26	22 24 34	22 25 40	22 26 19	22 27 15	22 28 10	22 29 18	22 30 34	22 31 13	22 32 12	22 33 47
22 34 31	22 35 21	22 36 24	22 37 36	22 38 22	22 39 45	22 40 63	22 41 51	22 42 51	22 43 38	22 44 44	22 45 47
22 46 21	22 47 30	22 48 17	22 49 30	22 50 24	22 51 21	23 1 22	23 2 34	23 3 37	23 4 31	23 5 36	23 6 11
23 7 6	23 8 16	23 9 43	23 10 50	23 11 31	23 12 29	23 13 34	23 14 16	23 15 46	23 16 39	23 17 36	23 18 24
23 19 44	23 20 41	23 21 48	23 22 26	23 23 0	23 24 9	23 25 21	23 26 16	23 27 17	23 28 29	23 29 43	23 30 52
23 31 24	23 32 25	23 33 56	23 34 51	23 35 46	23 36 49	23 37 38	23 38 36	23 39 60	23 40 52	23 41 40	23 42 47
23 43 13	23 44 44	23 45 52	23 46 24	23 47 27	23 48 9	23 49 43	23 50 45	23 51 22	24 1 29	24 2 41	24 3 46
24 4 29	24 5 39	24 6 14	24 7 14	24 8 25	24 9 48	24 10 53	24 11 36	24 12 30	24 13 27	24 14 11	24 15 46
24 16 45	24 17 35	24 18 21	24 19 39	24 20 49	24 21 55	24 22 34	24 23 9	24 24 0	24 25 14	24 26 25	24 27 22
24 28 38	24 29 50	24 30 56	24 31 34	24 32 31	24 33 57	24 34 56	24 35 55	24 36 58	24 37 38	24 38 41	24 39 63
24 40 46	24 41 35	24 42 44	24 43 12	24 44 43	24 45 52	24 46 27	24 47 26	24 48 17	24 49 47	24 50 50	24 51 25
25 1 33	25 2 43	25 3 52	25 4 18	25 5 34	25 6 17	25 7 27	25 8 34	25 9 45	25 10 47	25 11 35	25 12 25
25 13 13	25 14 6	25 15 36	25 16 45	25 17 25	25 18 11	25 19 26	25 20 54	25 21 55	25 22 40	25 23 21	25 24 14
25 25 0	25 26 36	25 27 25	25 28 46	25 29 52	25 30 52	25 31 43	25 32 32	25 33 48	25 34 54	25 35 60	25 36 64
25 37 30	25 38 38	25 39 57	25 40 32	25 41 21	25 42 31	25 43 26	25 44 33	25 45 43	25 46 25	25 47 19	25 48 25
25 49 42	25 50 49	25 51 23	26 1 19	26 2 29	26 3 25	26 4 43	26 5 40	26 6 22	26 7 11	26 8 7	26 9 43
26 10 53	26 11 31	26 12 36	26 13 48	26 14 30	26 15 53	26 16 37	26 17 45	26 18 36	26 19 57	26 20 32	26 21 44
26 22 19	26 23 16	26 24 25	26 25 36	26 26 0	26 27 20	26 28 16	26 29 37	26 30 51	26 31 10	26 32 25	26 33 61
26 34 49	26 35 35	26 36 36	26 37 46	26 38 38	26 39 62	26 40 66	26 41 54	26 42 58	26 43 22	26 44 53	26 45 59
26 46 29	26 47 36	26 48 13	26 49 45	26 50 42	26 51 28	27 1 8	27 2 19	27 3 27	27 4 24	27 5 21	27 6 9
27 7 20	27 8 14	27 9 27	27 10 34	27 11 14	27 12 16	27 13 34	27 14 19	27 15 33	27 16 23	27 17 25	27 18 20
27 19 39	27 20 29	27 21 33	27 22 13	27 23 17	27 24 22	27 25 25	27 26 20	27 27 0	27 28 23	27 29 28	27 30 35
27 31 22	27 32 8	27 33 41	27 34 34	27 35 35	27 36 39	27 37 26	27 38 20	27 39 44	27 40 49	27 41 37	27 42 39
27 43 30	27 44 33	27 45 39	27 46 9	27 47 17	27 48 9	27 49 27	27 50 28	27 51 8	28 1 16	28 2 19	28 3 9
28 4 47	28 5 37	28 6 30	28 7 26	28 8 13	28 9 35	28 10 47	28 11 26	28 12 37	28 13 57	28 14 40	28 15 51
28 16 28	28 17 47	28 18 43	28 19 62	28 20 17	28 21 31	28 22 10	28 23 29	28 24 38	28 25 46	28 26 16	28 27 23
28 28 0	28 29 24	28 30 43	28 31 6	28 32 22	28 33 57	28 34 38	28 35 19	28 36 20	28 37 46	28 38 32	28 39 54
28 40 72	28 41 60	28 42 61	28 43 38	28 44 54	28 45 57	28 46 30	28 47 39	28 48 22	28 49 39	28 50 33	28 51 30
29 1 21	29 2 9	29 3 17	29 4 44	29 5 25	29 6 37	29 7 44	29 8 30	29 9 16	29 10 28	29 11 17	29 12 31
29 13 58	29 14 46	29 15 39	29 16 9	29 17 40	29 18 44	29 19 57	29 20 10	29 21 7	29 22 18	29 23 43	29 24 50
29 25 52	29 26 37	29 27 28	29 28 24	29 29 0	29 30 21	29 31 30	29 32 20	29 33 40	29 34 15	29 35 16	29 36 22
29 37 37	29 38 18	29 39 33	29 40 68	29 41 57	29 42 53	29 43 55	29 44 43	29 45 42	29 46 28	29 47 37	29 48 34
29 49 22	29 50 11	29 51 29	30 1 33	30 2 24	30 3 37	30 4 38	30 5 18	30 6 42	30 7 55	30 8 44	30 9 8
30 10 9	30 11 21	30 12 27	30 13 53	30 14 48	30 15 25	30 16 15	30 17 31	30 18 41	30 19 47	30 20 31	30 21 16

210 Smallest Position Value Approach

30 22 34	30 23 52	30 24 56	30 25 52	30 26 51	30 27 35	30 28 43	30 29 21	30 30 0	30 31 47	30 32 28	30 33 21
30 34 7	30 35 36	30 36 42	30 37 26	30 38 15	30 39 12	30 40 57	30 41 49	30 42 41	30 43 65	30 44 30	30 45 25
30 46 29	30 47 33	30 48 43	30 49 10	30 50 10	30 51 31	31 1 17	31 2 23	31 3 16	31 4 46	31 5 39	31 6 27
31 7 21	31 8 9	31 9 39	31 10 50	31 11 28	31 12 37	31 13 54	31 14 37	31 15 53	31 16 32	31 17 47	31 18 41
31 19 61	31 20 23	31 21 37	31 22 13	31 23 24	31 24 34	31 25 43	31 26 10	31 27 22	31 28 6	31 29 30	31 30 47
31 31 0	31 32 23	31 33 60	31 34 43	31 35 26	31 36 26	31 37 47	31 38 35	31 39 58	31 40 71	31 41 59	31 42 61
31 43 32	31 44 54	31 45 59	31 46 30	31 47 39	31 48 18	31 49 42	31 50 37	31 51 30	32 1 6	32 2 11	32 3 23
32 4 27	32 5 16	32 6 17	32 7 27	32 8 17	32 9 19	32 10 28	32 11 6	32 12 16	32 13 39	32 14 26	32 15 30
32 16 15	32 17 25	32 18 25	32 19 41	32 20 22	32 21 24	32 22 12	32 23 25	32 24 31	32 25 32	32 26 25	32 27 8
32 28 22	32 29 20	32 30 28	32 31 23	32 32 0	32 33 37	32 34 26	32 35 29	32 36 34	32 37 25	32 38 13	32 39 37
32 40 52	32 41 40	32 42 40	32 43 38	32 44 32	32 45 36	32 46 9	32 47 19	32 48 16	32 49 21	32 50 20	32 51 10
33 1 43	33 2 39	33 3 54	33 4 31	33 5 21	33 6 45	33 7 60	33 8 54	33 9 24	33 10 12	33 11 31	33 12 27
33 13 44	33 14 47	33 15 12	33 16 32	33 17 23	33 18 37	33 19 33	33 20 49	33 21 36	33 22 47	33 23 56	33 24 57
33 25 48	33 26 61	33 27 41	33 28 57	33 29 40	33 30 21	33 31 60	33 32 37	33 33 0	33 34 27	33 35 55	33 36 61
33 37 18	33 38 25	33 39 14	33 40 41	33 41 37	33 42 25	33 43 68	33 44 17	33 45 7	33 46 32	33 47 30	33 48 50
33 49 18	33 50 29	33 51 34	34 1 31	34 2 20	34 3 32	34 4 42	34 5 21	34 6 42	34 7 53	34 8 42	34 9 9
34 10 16	34 11 21	34 12 30	34 13 57	34 14 50	34 15 30	34 16 12	34 17 35	34 18 44	34 19 52	34 20 25	34 21 9
34 22 31	34 23 51	34 24 56	34 25 54	34 26 49	34 27 34	34 28 38	34 29 15	34 30 7	34 31 43	34 32 26	34 33 27
34 34 0	34 35 30	34 36 36	34 37 31	34 38 16	34 39 18	34 40 62	34 41 53	34 42 46	34 43 64	34 44 36	34 45 32
34 46 30	34 47 36	34 48 42	34 49 14	34 50 6	34 51 32	35 1 27	35 2 19	35 3 10	35 4 56	35 5 40	35 6 44
35 7 45	35 8 31	35 9 32	35 10 43	35 11 30	35 12 44	35 13 69	35 14 54	35 15 54	35 16 24	35 17 53	35 18 54
35 19 70	35 20 7	35 21 21	35 22 21	35 23 46	35 24 55	35 25 60	35 26 35	35 27 35	35 28 19	35 29 16	35 30 36
35 31 26	35 32 29	35 33 55	35 34 30	35 35 0	35 36 6	35 37 51	35 38 33	35 39 48	35 40 81	35 41 69	35 42 67
35 43 57	35 44 58	35 45 58	35 46 38	35 47 48	35 48 38	35 49 38	35 50 27	35 51 39	36 1 31	36 2 24	36 3 12
36 4 61	36 5 45	36 6 47	36 7 46	36 8 33	36 9 38	36 10 49	36 11 35	36 12 49	36 13 73	36 14 58	36 15 59
36 16 30	36 17 58	36 18 58	36 19 75	36 20 13	36 21 27	36 22 24	36 23 49	36 24 58	36 25 64	36 26 36	36 27 39
36 28 20	36 29 22	36 30 42	36 31 26	36 32 34	36 33 61	36 34 36	36 35 6	36 36 0	36 37 56	36 38 38	36 39 54
36 40 86	36 41 74	36 42 72	36 43 58	36 44 63	36 45 64	36 46 43	36 47 53	36 48 40	36 49 44	36 50 33	36 51 44
37 1 30	37 2 32	37 3 47	37 4 13	37 5 11	37 6 27	37 7 44	37 8 40	37 9 23	37 10 19	37 11 21	37 12 10
37 13 27	37 14 28	37 15 7	37 16 28	37 17 5	37 18 19	37 19 21	37 20 44	37 21 36	37 22 36	37 23 38	37 24 38
37 25 30	37 26 46	37 27 26	37 28 46	37 29 37	37 30 26	37 31 47	37 32 25	37 33 18	37 34 31	37 35 51	37 36 56
37 37 0	37 38 18	37 39 28	37 40 31	37 41 23	37 42 16	37 43 50	37 44 7	37 45 14	37 46 17	37 47 12	37 48 34
37 49 17	37 50 28	37 51 18	38 1 19	38 2 15	38 3 30	38 4 27	38 5 7	38 6 27	38 7 40	38 8 30	38 9 7
38 10 15	38 11 7	38 12 14	38 13 41	38 14 34	38 15 21	38 16 9	38 17 22	38 18 28	38 19 39	38 20 26	38 21 18
38 22 22	38 23 36	38 24 41	38 25 38	38 26 38	38 27 20	38 28 32	38 29 18	38 30 15	38 31 35	38 32 13	38 33 25
38 34 16	38 35 33	38 36 38	38 37 18	38 38 0	38 39 24	38 40 49	38 41 39	38 42 35	38 43 49	38 44 25	38 45 26
38 46 14	38 47 20	38 48 28	38 49 8	38 50 11	38 51 16	39 1 43	39 2 35	39 3 49	39 4 41	39 5 24	39 6 50
39 7 64	39 8 55	39 9 19	39 10 10	39 11 31	39 12 33	39 13 55	39 14 54	39 15 23	39 16 27	39 17 33	39 18 46
39 19 46	39 20 43	39 21 27	39 22 45	39 23 60	39 24 63	39 25 57	39 26 62	39 27 44	39 28 54	39 29 33	39 30 12
39 31 58	39 32 37	39 33 14	39 34 18	39 35 48	39 36 54	39 37 28	39 38 24	39 39 0	39 40 55	39 41 49	39 42 38
39 43 73	39 44 29	39 45 21	39 46 36	39 47 38	39 48 52	39 49 17	39 50 22	39 51 38	40 1 56	40 2 62	40 3 75
40 4 25	40 5 42	40 6 44	40 7 58	40 8 62	40 9 54	40 10 48	40 11 51	40 12 37	40 13 19	40 14 37	40 15 33
40 16 59	40 17 28	40 18 30	40 19 11	40 20 74	40 21 67	40 22 63	40 23 52	40 24 46	40 25 32	40 26 66	40 27 49
40 28 72	40 29 68	40 30 57	40 31 71	40 32 52	40 33 41	40 34 62	40 35 81	40 36 86	40 37 31	40 38 49	40 39 55
40 40 0	40 41 12	40 42 16	40 43 58	40 44 27	40 45 34	40 46 43	40 47 33	40 48 53	40 49 48	40 50 60	40 51 42
41 1 44	41 2 50	41 3 63	41 4 13	41 5 33	41 6 32	41 7 47	41 8 50	41 9 45	41 10 41	41 11 40	41 12 26
41 13 9	41 14 25	41 15 26	41 16 48	41 17 18	41 18 17	41 19 5	41 20 62	41 21 58	41 22 51	41 23 40	41 24 35
41 25 21	41 26 54	41 27 37	41 28 60	41 29 57	41 30 49	41 31 59	41 32 40	41 33 37	41 34 53	41 35 69	41 36 74
41 37 23	41 38 39	41 39 49	41 40 12	41 41 0	41 42 13	41 43 47	41 44 20	41 45 30	41 46 31	41 47 21	41 48 41
41 49 40	41 50 50	41 51 30	42 1 45	42 2 48	42 3 62	42 4 16	42 5 28	42 6 37	42 7 53	42 8 53	42 9 39
42 10 32	42 11 37	42 12 24	42 13 22	42 14 33	42 15 16	42 16 44	42 17 14	42 18 23	42 19 9	42 20 60	42 21 52
42 22 51	42 23 47	42 24 44	42 25 31	42 26 58	42 27 39	42 28 61	42 29 53	42 30 41	42 31 61	42 32 30	42 33 25
42 34 46	42 35 67	42 36 72	42 37 16	42 38 35	42 39 38	42 40 16	42 41 13	42 42 0	42 43 56	42 44 10	42 45 18
42 46 31	42 47 22	42 48 45	42 49 32	42 50 44	42 51 31	43 1 34	43 2 46	43 3 47	43 4 41	43 5 49	43 6 23
43 7 12	43 8 26	43 9 56	43 10 63	43 11 44	43 12 41	43 13 39	43 14 23	43 15 57	43 16 52	43 17 47	43 18 33
43 19 52	43 20 52	43 21 61	43 22 38	43 23 13	43 24 12	43 25 26	43 26 22	43 27 30	43 28 38	43 29 55	43 30 65
43 31 32	43 32 38	43 33 68	43 34 64	43 35 57	43 36 59	43 37 50	43 38 49	43 39 73	43 40 58	43 41 47	43 42 56
43 43 0	43 44 55	43 45 64	43 46 37	43 47 38	43 48 22	43 49 56	43 50 58	43 51 35	44 1 38	44 2 39	44 3 54
44 4 15	44 5 18	44 6 33	44 7 50	44 8 47	44 9 28	44 10 22	44 11 29	44 12 17	44 13 27	44 14 32	44 15 6
44 16 34	44 17 9	44 18 22	44 19 17	44 20 51	44 21 42	44 22 44	44 23 44	44 24 43	44 25 33	44 26 53	44 27 33
44 28 54	44 29 43	44 30 30	44 31 54	44 32 32	44 33 17	44 34 36	44 35 58	44 36 63	44 37 7	44 38 25	44 39 29
44 40 27	44 41 20	44 42 10	44 43 55	44 44 0	44 45 10	44 46 24	44 47 18	44 48 40	44 49 22	44 50 34	44 51 25
45 1 42	45 2 40	45 3 56	45 4 25	45 5 20	45 6 41	45 7 57	45 8 53	45 9 26	45 10 16	45 11 31	45 12 23
45 13 37	45 14 42	45 15 7	45 16 34	45 17 18	45 18 32	45 19 26	45 20 51	45 21 39	45 22 47	45 23 52	45 24 52

Smallest Position Value Approach

45 25 43	45 26 59	45 27 39	45 28 57	45 29 42	45 30 25	45 31 59	45 32 36	45 33 7	45 34 32	45 35 58	45 36 64
45 37 14	45 38 26	45 39 21	45 40 34	45 41 30	45 42 18	45 43 64	45 44 10	45 45 0	45 46 30	45 47 26	45 48 47
45 49 20	45 50 32	45 51 31	46 1 14	46 2 20	46 3 32	46 4 18	46 5 12	46 6 14	46 7 28	46 8 23	46 9 21
46 10 26	46 11 10	46 12 7	46 13 30	46 14 20	46 15 23	46 16 20	46 17 17	46 18 16	46 19 32	46 20 31	46 21 30
46 22 21	46 23 24	46 24 27	46 25 25	46 26 29	46 27 9	46 28 30	46 29 28	46 30 29	46 31 30	46 32 9	46 33 32
46 34 30	46 35 38	46 36 43	46 37 17	46 38 14	46 39 36	46 40 43	46 41 31	46 42 31	46 43 37	46 44 24	46 45 30
46 46 0	46 47 10	46 48 17	46 49 19	46 50 24	46 51 2	47 1 23	47 2 29	47 3 42	47 4 8	47 5 15	47 6 16
47 7 32	47 8 31	47 9 27	47 10 28	47 11 19	47 12 6	47 13 21	47 14 16	47 15 19	47 16 28	47 17 9	47 18 8
47 19 22	47 20 41	47 21 38	47 22 30	47 23 27	47 24 26	47 25 19	47 26 36	47 27 17	47 28 39	47 29 37	47 30 33
47 31 39	47 32 19	47 33 30	47 34 36	47 35 48	47 36 53	47 37 12	47 38 20	47 39 38	47 40 33	47 41 21	47 42 22
47 43 38	47 44 18	47 45 26	47 46 10	47 47 0	47 48 23	47 49 23	47 50 31	47 51 9	48 1 12	48 2 25	48 3 28
48 4 29	48 5 29	48 6 9	48 7 11	48 8 9	48 9 35	48 10 43	48 11 22	48 12 24	48 13 36	48 14 18	48 15 41
48 16 30	48 17 32	48 18 23	48 19 44	48 20 32	48 21 39	48 22 17	48 23 9	48 24 17	48 25 25	48 26 13	48 27 9
48 28 22	48 29 34	48 30 43	48 31 18	48 32 16	48 33 50	48 34 42	48 35 38	48 36 40	48 37 34	48 38 28	48 39 52
48 40 53	48 41 41	48 42 45	48 43 22	48 44 40	48 45 47	48 46 17	48 47 23	48 48 0	48 49 35	48 50 36	48 51 16
49 1 26	49 2 21	49 3 36	49 4 28	49 5 8	49 6 33	49 7 47	49 8 38	49 9 6	49 10 8	49 11 14	49 12 17
49 13 43	49 14 39	49 15 17	49 16 14	49 17 22	49 18 31	49 19 38	49 20 31	49 21 20	49 22 30	49 23 43	49 24 47
49 25 42	49 26 45	49 27 27	49 28 39	49 29 22	49 30 10	49 31 42	49 32 21	49 33 18	49 34 14	49 35 38	49 36 44
49 37 17	49 38 8	49 39 17	49 40 48	49 41 40	49 42 32	49 43 56	49 44 22	49 45 20	49 46 19	49 47 23	49 48 35
49 49 0	49 50 12	49 51 22	50 1 24	50 2 14	50 3 27	50 4 38	50 5 17	50 6 36	50 7 47	50 8 35	50 9 6
50 10 17	50 11 15	50 12 25	50 13 52	50 14 44	50 15 29	50 16 6	50 17 32	50 18 39	50 19 49	50 20 21	50 21 8
50 22 24	50 23 45	50 24 50	50 25 49	50 26 42	50 27 28	50 28 33	50 29 11	50 30 10	50 31 37	50 32 20	50 33 29
50 34 6	50 35 27	50 36 33	50 37 28	50 38 11	50 39 22	50 40 60	50 41 50	50 42 44	50 43 58	50 44 34	50 45 32
50 46 24	50 47 31	50 48 36	50 49 12	50 50 0	50 51 26	51 1 14	51 2 21	51 3 33	51 4 17	51 5 14	51 6 11
51 7 26	51 8 22	51 9 23	51 10 28	51 11 12	51 12 8	51 13 29	51 14 18	51 15 25	51 16 22	51 17 17	51 18 15
51 19 32	51 20 32	51 21 32	51 22 21	51 23 22	51 24 25	51 25 23	51 26 28	51 27 8	51 28 30	51 29 29	51 30 31
51 31 30	51 32 10	51 33 34	51 34 32	51 35 39	51 36 44	51 37 18	51 38 16	51 39 38	51 40 42	51 41 30	51 42 31
51 43 35	51 44 25	51 45 31	51 46 2	51 47 9	51 48 16	51 49 22	51 50 26	51 51 0			

Author Index

Chen, Angela   121

Davendra, Donald   1, 13, 35

Gattoufi, Said   121
Gencyilmaz, Gunes   121

Liang, Yun-Chia   139
Lichtblau, Daniel   81

Onwubolu, Godfrey   1, 13, 35

Pan, Quan-Ke   139

Suganthan, Ponnuthurai   139

Tasgetiren, Fatih   121, 139

Zelinka, Ivan   163

Printing: Krips bv, Meppel, The Netherlands
Binding: Stürtz, Würzburg, Germany